Logic Design Projects Using Standard Integrated Circuits

Logic Design Projects Using Standard Integrated Circuits

JOHN F. WAKERLY
Stanford University

JOHN WILEY & SONS
New York   Santa Barbara   London   Sydney   Toronto

LIBRARY OF CONGRESS CATALOGING IN PUBLICATION DATA:

Wakerly, John.
Logic design projects using standard integrated circuits.

Includes bibliographical references.
1. Digital electronics--Laboratory manuals. 2. Logic design--Laboratory manuals. 3. Integrated circuits--Laboratory manuals. I. Title.
TK7868.D5W3 621.3819'58'2 76-5471
ISBN 0-471-91705-2

Printed in the United States of America
10 9 8 7 6 5

to Ralph and Carm

Preface

This manual is a sourcebook for design projects for college-level digital logic and microprocessor laboratories. Projects include introductory experiments that can be performed in conjunction with a standard lecture course on logic design, complex projects that can be undertaken in an advanced lab course following a logic design lecture sequence, and microprocessor projects. The projects are designed and constructed using standard, inexpensive, reusable integrated circuits and plug-in breadboards that are available from many manufacturers. In addition to the project descriptions, the lab manual contains three chapters of narrative material that must be provided in some form or another in any digital lab course. Hard-to-find background material is also provided with many of the project descriptions.

Projects are ordered according to difficulty, with a unit of one week corresponding to the effort expended by a typical college senior in one week of a 3 credit-hour course. The manual contains 35 projects with difficulties of one to four weeks each, for a total of 69 weeks. In addition there are 13 microprocessor projects. Thus the instructor and students have considerable flexibility in tailoring individual programs according to interest and equipment availability.

We assume that students who use this manual will be taking or will have taken a logic design course that employs a standard text. This manual makes no attempt to be a substitute for a good text. However, we do include some tutorial material for the A-series introductory projects and some background material not normally covered in typical logic design lecture courses.

Chapter One describes the common features of typical logic breadboards and provides hints for using them. Chapter Two gives standards for project documentation. Familiarity with this material will help the student not only in the logic lab but also in the environment of industrial logic design. Chapter Three presents several guidelines for project design and debugging that will make it easier for students to complete their projects successfully.

Chapter Four contains eleven introductory projects. Each of these projects is designed to give the student practical experience with a particular device or concept from his lecture course. These projects are most suitable for introductory lab courses taken concurrently with lectures.

Chapter Five through Eight contain projects directed toward designing, building, and debugging a particular system rather than learning a specific concept. The resulting systems are interesting and often fun to operate - consequently the student is motivated to carry out these projects to successful completion. In each project the student is exposed to a number of different concepts that he "learns by doing." There are 6 such one-week projects, 7 two-week projects, 6 three-week projects, and 5 four-week projects.

Chapter Nine contains both introductory and advanced microprocessor projects. Except for a set of introductory projects for three particular microprocessors, all of the projects can be used with any microprocessor. These projects are suitable for a microprocessor lab that begins with simple projects and ends with a three- or four-week final project. Many of the projects are simply microprocessor versions of projects in Chapters Four through Eight, so the student may have the opportunity to compare hard-wired and microprocessor approaches to the same problem.

The material in this manual was created during a two-year lab development effort in the Electrical Engineering Department at Stanford University. Much credit goes to Professors Edward J. McCluskey, Ralph Smith, and John Linvill, who saw the need for and sponsored the development effort. Equipment and some salary support for our digital logic laboratory was provided by a series of Graduate Grants from the General Electric Foundation, and gifts of ICs that made the advanced projects possible were received from Intel Corporation and Texas Instruments, Inc. Equipment for our microprocessor lab came from a substantial gift from Intel Corporation and grants from the General Electric Foundation, the National Science Foundation, and Motorola. We are very grateful for this help.

A substantial number of people contributed project ideas and other material to the manual. The most substantial contribution was made by Jim McClure who not only produced many project ideas and descriptions, but also contributed the ideas and experience behind Chapters Two and Three. Thanks are also due to Mark Hahn who invented half a dozen projects, and to Professor Ed Davidson, who began the whole logic lab program six years ago. Dr. Vic Grinich aided greatly in the development of microprocessor projects. Many others contributed project ideas, including Bruce Almich, Dag Belsnes, Mike Carter, Dominic Chirieleison, Bruce Eisenhard, Moe Rubenzahl, Fred Sammartino, and Terry Walker.

As usual, thanks go to all the typists who worked on the preliminary manuscript, including Moira Lieberman, Vicky Gahart, Liz Laughead, Naomi Schulman, and myself. Special thanks and credit go to Moira Lieberman for her fine typing of the final camera-ready manuscript.

Palo Alto, California *John F. Wakerly*

Preface to the Instructor

The purpose of this manual is to provide a framework for experimentation with digital electronics. It is not a self-study course in introductory logic design but rather a sourcebook of project ideas and tidbits of information that can turn a student with some digital lecture exposure into a proficient logic designer. Thus, throughout the manual we assume that the reader is well grounded in the logic design fundamentals taught in typical lecture courses, but we emphasize concepts, subtleties, and tricks that can only be learned by practical experience.

DIGITAL LOGIC LAB FORMAT

The projects in Chapters Four through Eight of this manual are rated according to difficulty, where the basic unit of difficulty is one week. One week corresponds to the effort that a typical college senior devotes to a three-unit course in one week. At Stanford we use the rule-of-thumb that a student should devote three hours per unit per week to each class, including lectures, home study, and labs; so nine hours per week should be devoted to a three-unit course. Student surveys taken in the three-unit Stanford class using this manual indicate that the average student spends seven to eight hours per week, including a weekly one-hour lecture. The typical breakdown of a student's time is three hours at home for design and reports, three hours in the lab, and one hour at the lecture per week.

Since there are far more projects in the manual than any single student will undertake, we have found that it is most desirable to give each student the maximum possible freedom to choose his own laboratory program. Thus, for a 10-week academic quarter, we ask each student to devise his own unique program of projects so that the total of his projects' ratings is 9 or 10 weeks. Of course, we must constrain the students according to the availability of parts, and we must occassionally adjust the program of students who have chosen projects much too easy or much too difficult for their ability. Students may repeat the class in a later quarter by choosing a different set of projects.

When this manual is used in conjunction with an introductory logic design lecture course, only the A-series introductory one-week projects should be assigned at the beginning of the course. The first seven A-series projects correspond to standard lecture material, and the last four projects can be used as desired. At the end of a typical lecture course, the student should be ready to handle a B-series or C-series project as the final assignment.

At Stanford we require our lab students to have already taken the logic design course as a prerequisite, so that they will have all of the theoretical background necessary to start on the lab. We advise these students to begin with one or two A-series projects to familiarize themselves with the equipment, but then we require them to move on to a B-series and more difficult projects. Even with no previous practical experience, most students can successfully undertake a D-series or E-series project by the end of the quarter.

Our laboratory situation at Stanford allows each student to have an individual logic breadboard and free access to the laboratory at any time to work on projects. However, it is possible for students to completely build and debug one-week projects during a scheduled four-hour laboratory session, *if* they have done their initial designs before the lab session begins. It is likewise possible to do longer projects in a series of scheduled sessions if each student's breadboard setup can remain intact from session to session.

As mentioned earlier, a weekly one-hour lecture is included in the lab course at Stanford. The first three or four lectures cover administrative procedures and material from Chapters One through Three. Remaining lectures are used for sample project solutions and for topics of interest not covered in other lecture courses. Past topics have included the following:

TTL Characteristics

LEDs and 7-segment Displays

Digital-to-Analog and Analog-to-Digital Converters

Peripheral Interfacing

Serial Input/Output

Transmission Lines and Cabling

ECL and Schottky TTL High Speed Logic

MICROPROCESSOR LAB FORMAT

The microprocessor projects in Chapter Nine are designed for a lab that begins with simple introductory projects and ends with a three- or four-week final project. The lab should begin with Project M1, M2, or M3, depending on the microprocessor used. In a three-unit, ten-week lab at Stanford, we make Assignments I and II of the project due after the first two weeks of the lab, Assignments III and IV after the third week, and Assignment V after the fourth week. The student spends two weeks on Project M6, and then he is free to choose a four-week final project. The complexity of the microprocessor projects varies for the different projects, and even the difficulty of a particular project may depend on both the approach and equipment availability. As an instructor you must determine how much time to allocate according to your own situation.

In the introductory project (M1, M2, or M3) each student builds his own small microprocessor system using a logic breadboard kit. The remaining projects can be built with the breadboard alone or with a combination of the breadboard and a manufacturer's microcomputer prototyping system.

It will almost certainly be necessary to accompany the lab with a series of lectures to introduce the student to the microprocessor being used. At Stanford we assume that the student has had introductory courses in digital logic design and some minicomputer assembly language as prerequisites, and that he has had some practical experience with TTL logic (for example, a digital logic lab). These prerequisites allow us to cover all the student needs to know about our particular microprocessor, the 8080, in nine or ten lectures (two one-hour lectures per week for five weeks). The following topics are covered (about one lecture each).

8080 Architecture and Instruction Set

8080 Assembler and more on the Instruction Set

MDS Microcomputer Development System

8080 Electrical Specifications

8080 Support Chips

Design of a small 8080 System

Interrupts and DMA in the 8080

Serial I/O and UARTs.

PL/M

For the second half of the quarter we reduce to one lecture per week, using the time to discuss other microprocessors. Aside from this discussion, the continued weekly lectures are necessary primarily for a regular point of contact for making announcements, answering student's questions, and so on.

TEXTS AND EQUIPMENT

The required text for use with the lab manual is a manufacturer's data book for the integrated circuits used in the lab. For TTL projects the most readily available and accurate book is *The TTL Data Book for Design Engineers*, available from the Texas Instruments Learning Center, P.O. Box 5012, Dallas, TX 75222. TTL data books are also available from National Semiconductor (2900 Semiconductor Drive, Santa Clara, CA 95051), Signetics (811 E. Arques Ave., Sunnyvale, CA 94086), and Fairchild (464 Ellis St., Mountain View, CA 94042). CMOS data books can also be obtained from the manufacturers, including those above, RCA (P.O. Box 3200, Somerville, NJ 08876), and Motorola (P.O. Box 20912, Phoenix, AZ 85036). For projects that use non-7400 parts like memories and DACs, it is best to make copies of just those data sheets necessary and hand them out to the students. Microprocessor literature should be obtained from the manufacturer of the particular microprocessor used in your lab. Examples of useful literature are *8080 System Users Manual* from Intel, *M6800 Systems Reference and Data Sheets* from Motorola, and *2650 Microprocessor*, from Signetics.

The projects in the manual are small; all of them can be fit on three standard plug-in breadboard strips of the type described in Chapter One. In order to perform the projects, a student should have three breadboards strips, a 5 volt power supply, some switches for inputs, and some lamps for output. All of the projects can be done with only eight toggle switches (undebounced), four push buttons (debounced), and twelve lamps. Modifications can be made to the project descriptions when fewer lamps and switches are available. Individual breadboard strips or complete breadboards with power supply, lamps, and switches are available from several manufacturers: E & L Instruments (61 First St., Derby, CT 06418), A P Products (P.O. Box 110-Q, Painesville, OH 44077), Continental Specialties (44 Kendall St., P.O. Box 1942, New Haven, CT 06509), Jermyn (712 Montgomery St., San Francisco, CA 94111), The Technical Education Press (P.O. Box 342, Seal Beach, CA 90740), Logic Design, Inc. (P.O. Box 3991 University Station, Laramie, WY 82071), and Hewlett-Packard (1501 Page Mill Road, Palo Alto, CA 94304). At Stanford we designed and built our own breadboards from standard components at a parts cost of about $200 each.

Each project description in Chapters Four through Eight gives a list of parts needed in addition to a "standard kit" of integrated circuits. The standard kit, listed in Appendix A, is a set of commonly used parts that will be sufficient for the implementation of smaller projects, and for the control section of larger projects. At Stanford we issue each student a standard kit along with his breadboard, and we require the students to check out additional parts for larger projects as they need them.

The project descriptions in Chapters Four through Eight are all based on standard 7400-series TTL or 74C00-series CMOS circuits. However, the projects can also be performed using 4000-series CMOS; the 4000-series functional replacements for 7400-series logic are indicated in Appendix B.

A digital logic lab can be run without any test equipment at all, but a few good instruments enhance the student's learning experience.

Only two basic instruments are needed: a pulse generator and an oscilloscope. The pulse generator should allow the user to set both the period and pulse width of the output and should generate TTL-compatible signals. The oscilloscope should be a dual-trace scope with a bandwidth of 25 MHz or better, and should preferably have delay lines in the vertical amplifiers so that the triggering edges of pulses can be observed; however, delayed sweep is not necessary. A separate power supply is necessary only if components such as DACs or MOS memories require it. A very useful and inexpensive instrument for debugging purposes is a three-state logic probe.

The only really special equipment is required by the microprocessor projects in Chapter Nine. Although the students can use standard breadboards to construct their microprocessor systems, they will still need a prototyping system such as Intel's Microcomputer Development System (MDS) to assemble and test their programs. A prototyping system is also useful if breadboarding facilities are limited since some projects, such as M4, M8, and M9, can be done completely on the prototyping system if it has lamps and switches. Other projects, such as M5, M6, M7, M10, M11, and M12, can use the prototyping system's processor and memory, implementing only special I/O functions on a breadboard. The above assumes that the prototyping system has general purpose I/O ports that can be accessed by the user via a cable.

Introductory projects are provided for the 8080 (manufactured by Intel, Advanced Micro Devices, and Texas Instruments), the 6800 (Motorola and American Microsystems, Inc.), and the 2650 (Signetics). Different basic system support chips such as clock drivers are required for each microprocessor. Different I/O chips are offered in each microprocessor family, but they can be classified into two general types. One type is the simple I/O latch, and as the Intel 8212 or Signetics 8T31. The other type is the programmable interface adapter (PIA) such as the Motorola 6820 or the Intel 8255. Both PROMs and RAMs will be needed. The 1702-type erasable PROMs are very convenient, of course. Byte-organized static RAMs such as the Motorola 6810 (128 × 8) are more inexpensive and convenient than using 1K RAMs such as 2102's when the memory requirements are small.

It is possible to debug simple microprocessor projects with only standard test equipment, but a logic state analyzer is a very useful tool. An in-circuit emulator, such as Intel's ICE-80, for a particular microprocessor is even more useful.

Contents

Logic Design Projects Using Standard Integrated Circuits

1. Logic Breadboards

Logic breadboards are available from a number of manufacturers, and the components for custom-built breadboards are also widely available (see Figure 1-1). Most of today's logic breadboards have a number of socket strips for plugging in integrated circuits (ICs) and other components, a power supply, lamps, and switches. More sophisticated models may also have built-in pulse generators, counters, and seven-segment displays. In this chapter we describe some of the basic features of logic breadboards and give hints for their effective use.

SOCKET STRIPS

Most socket strips are similar to Figure 1-2. The socket consists of an array of holes on 0.1 in. centers, corresponding to the standard pin spacing of ICs. Most of the wiring activity takes place in the center columns of the strip. On either side of the center gutter, the strip illustrated in Figure 1-2 has an array of 64 rows by 5 columns of socket terminals (holes). The five holes in each row are connected together electrically. The spacing across the gutter is 0.3 in., the same as the spacing between the rows of pins of a standard 14-, 16-, or 18-pin IC dual in-line package (DIP). Hence, an IC can be plugged in as shown in the figure, and for each IC pin there will be four holes remaining that are electrically connected to it. These holes are used for point-to-point interconnections to other IC pins, power, output lamps, and input switches. Larger ICs such as 24-pin DIPs have 0.6 in. row spacing; they can be used but they will

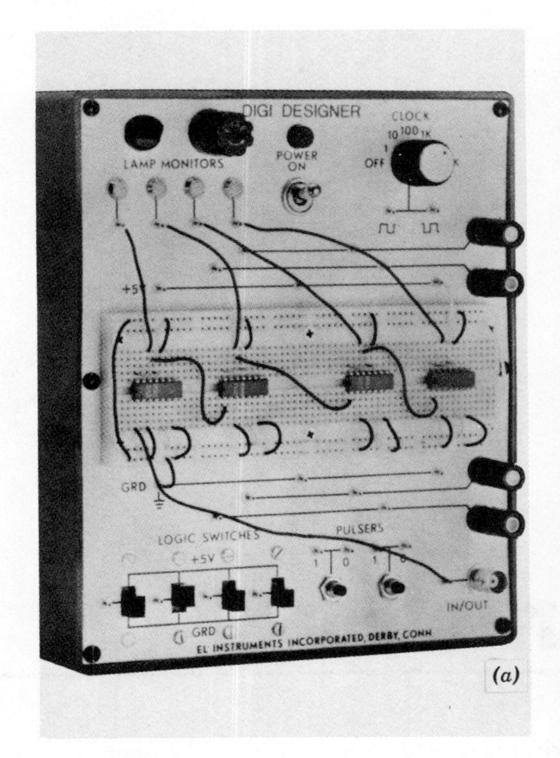

FIGURE 1-1 (a) Digi-Designer (courtesy of E&L Instruments); (b) Breadboard II (courtesy of A P Products).

leave fewer holes for interconnections. Discrete components such as resistors and transistors can also be plugged in.

The strip shown in the figure has two additional columns of holes at each edge of the strip. In this particular strip the columns are each divided in half so that there are in total eight sets of 25 electrically common holes. Such holes are usually used for the distribution of power and ground, clocks, and so on.

All plug-in strips accept standard IC leads and a variety of solid hookup wire sizes, usually from #20 to #26. It has been our experience and others' that #22 wire works best. In any case, the hookup wire must be solid, not stranded. The leads of discrete components such as switching transistors, bypass capacitors, ¼ watt resistors, and so on, can easily be inserted into the holes, but the leads of such components as larger filter capacitors or ½ watt resistors should not be forced into the holes.

INTEGRATED CIRCUITS AND WIRES

ICs are not difficult to insert in plug-in strips, once they have been properly conditioned. New ICs are shipped with the pins bent apart from the vertical to facilitate handling by automatic insertion equip-

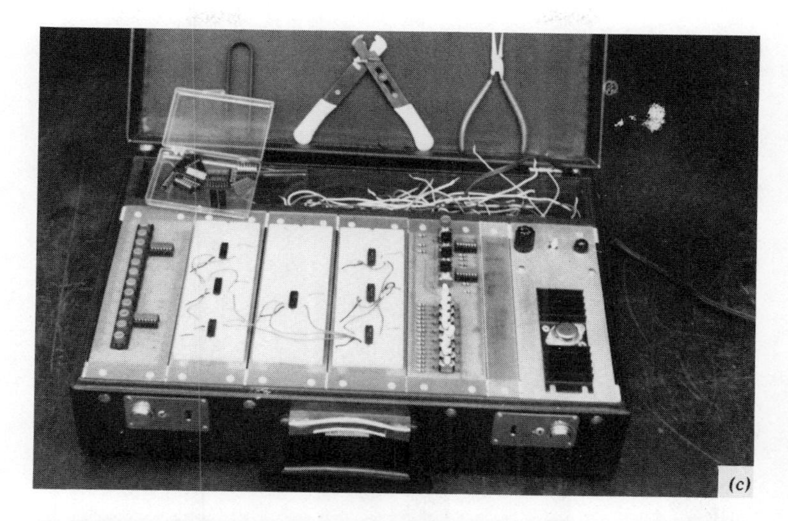

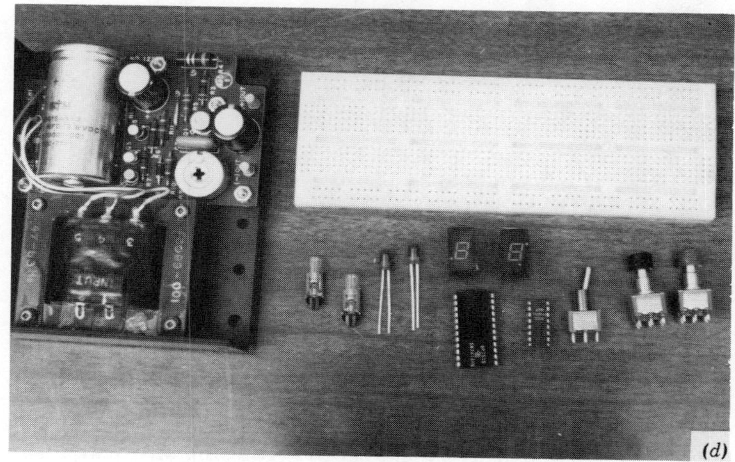

FIGURE 1-1 (c) Stanford University's custom-designed breadboards; (d) Components for custom breadboards.

ment. Therefore, before an IC is used for the first time the pins must be bent back so that their spacing is exactly 0.3 in. This can be done fairly well by squeezing the rows of pins between the thumb and forefinger, or very well by bending each row of pins at one time by grasping the entire row with a pair of long-nose pliers.

In order to protect expensive ICs from damage, your lab may mount them in sockets. If you have any such ICs, plug the socket directly into the breadboard. Never remove the IC from its socket -- the idea is to prevent IC pin damage by not using the IC pins.

An extraction tool should always be used for removing ICs. A very simple extraction tool is the U-shaped clip shown in Figure 1-3a. When ICs are plugged in tightly, attempts to remove them by hand usually result in bent pins and pain, as illustrated in Figure 1-3b.

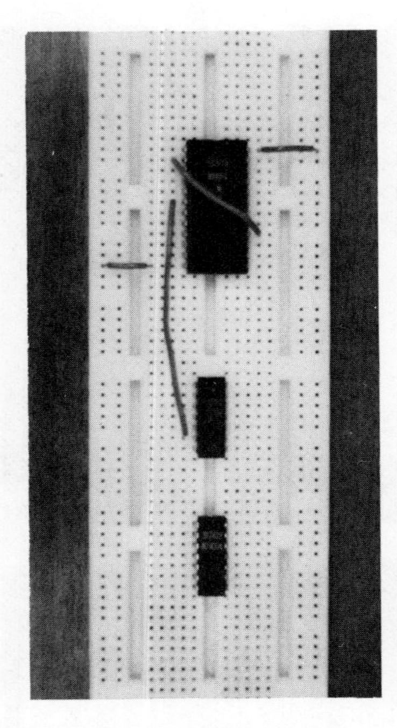

FIGURE 1-2 Breadboard socket strip.

All ICs should be inserted with the same orientation to facilitate wiring and debugging. It does *not* pay to reverse the orientation of some ICs to minimize wire lengths. The pin numberings of 14-, 16-, and 24-pin ICs are shown in Figure 1-4. A locating notch or hole identifies pin 1. In general, IC ground pins are on the left and power pins are on the right. Most breadboards also have ground buses on the left and power buses on the right. Hence it is best to insert *all* ICs with pin 1 in the upper left-hand corner (even strange ICs that have their power and ground pins in nonstandard locations). Double check the data book before wiring power and ground to any IC that you are not completely familiar with.

A pair of long-nose pliers is useful for inserting and removing wires from plug-in strips, especially when wiring becomes crowded. A pair of wire strippers is needed for cutting wires to length and removing about ½ in. of insulation from each end. Wires are very easy to insert, unless the stripped end is bent. Ends get bent after several uses of the same wire. A bent end is very difficult to insert, and if it is successfully inserted it could spread the terminal in the hole too much or break off in the hole. Ends that are not too severely bent should be straightened with the pliers, otherwise the end should be cut off and a new end stripped. Holding the wire stripper at a 45° angle when cutting a wire will produce a point that makes it easier to insert.

In general it is best to run all wires around ICs, not over them. This will make debugging easier and it will also make it easy to remove and replace an IC if you need to. Try not to cover up too many unused

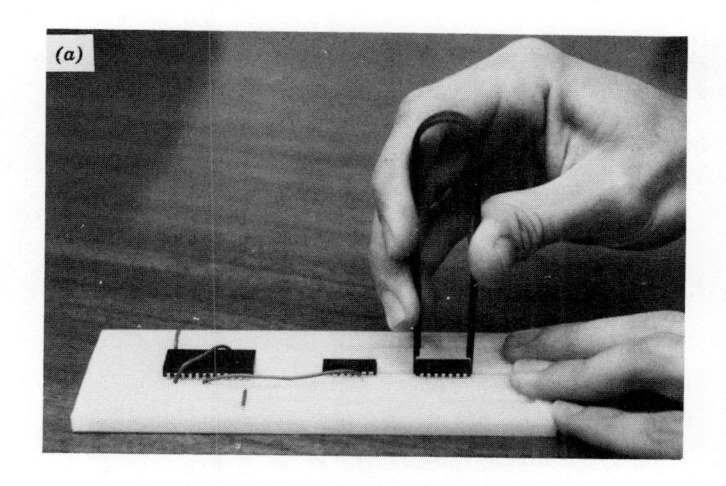

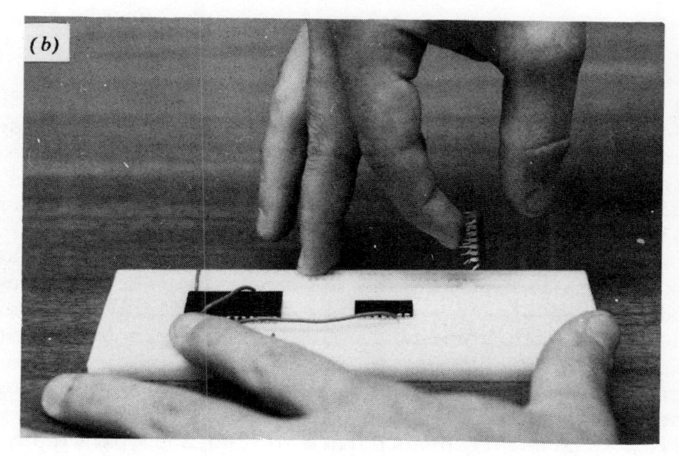

FIGURE 1-3 (a) IC extractor; (b) IC removal without an extractor.

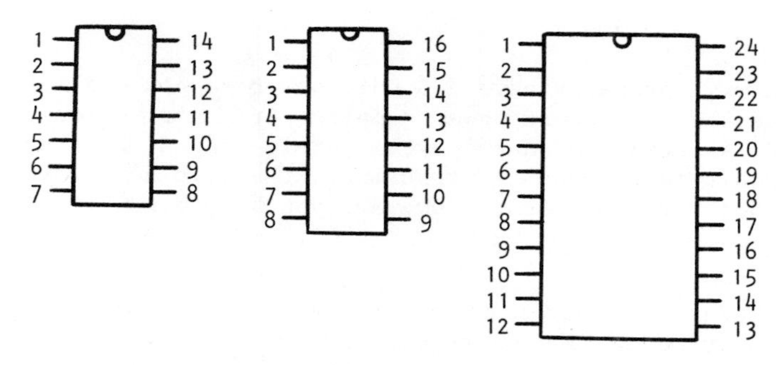

FIGURE 1-4 IC pin numberings.

holes with your wire runs. Keep wires close to the surface of the breadboard, and make them as short as possible subject to the preceding constraints.

Debugging your circuits will be easier if you follow a wiring color code for different types of signals. For example, the following color code could be used for microprocessor projects.

RED - +5 volts

BLACK - Ground

YELLOW - +12 volts

WHITE - Negative voltages

VIOLET - Control

ORANGE - Data bus

BROWN - Address bus

Wire the power to each IC on your breadboard first. Then wire unused inputs to an appropriate constant source, +5 volts (through a 1K current-limiting resistor) or ground. Next do all regular buses and then do control wiring. It should be apparent that this wiring order makes the first connections those that are least likely to change during debugging. When you are done wiring examine each pin of each IC sequentially and verify its connection -- this will save headaches later. Finally double-check all power connections before applying power to the circuit.

The importance of neat wiring cannot be overemphasized. Figure 1-5 shows examples of neat and messy wiring. Neat wiring jobs are easier to debug and are more reliable. Reliability is a very important consideration when project complexity involves over a dozen ICs. In a messy wiring job, removing or inserting one wire can have an unpredictable and often untraceable effect of removing (or worse, merely loosening) other wires tangled with it in the maze. It has been the policy at Stanford not to take off points for sloppy wiring, but to accept no excuses for project unreliability or inability to debug a project if the wiring is sloppy. In a neatly wired project, the teaching assistant or instructor at least has some chance of successfully helping the student debug.

POWER SUPPLIES

You will probably have a 5 volt power supply available with your breadboard. Modern power supplies are short-circuit protected, so that shorting the supply will merely shut it down until the short circuit is removed. Furthermore, there is no worry about getting a shock from 5 volts.* However, you *should* worry about blowing out

---

*A hazard does exist when working with supplies that can deliver 25 amps or more. The short-circuit current may be sufficient to vaporize the shorting wire or object. When working with circuits powered by such "monster" supplies, rings, watches, and other metal jewelry should be removed.

FIGURE 1-5 (a) Neat wiring; (b) Messy wiring.

ICs by improper application of power. The one sure way to permanently damage an IC is to reverse power and ground. Most ICs have their ground pin in the lower left-hand corner and power in the upper right-hand corner. Therefore plugging in an IC upside down will blow it out. Some ICs, most notably 7476, 7483, 7493, 7496, and 74100, have power and ground in nonstandard locations. Carelessness with these ICs can also be damaging.

Most of the time you will not damage an IC by shorting one of its outputs. TTL totem-pole outputs can be shorted together or to ground without damage. However, when an output trying to maintain a LOW level is shorted to the 5 volt supply there is usually damage. Shorting the outputs of a CMOS gate to each other or to power or ground for a short period of time is not damaging, but permanent damage or reduced device life could result after a few minutes of such a short. CMOS ICs have one unique problem. If a drive signal is applied to the input of a device that does not have power and

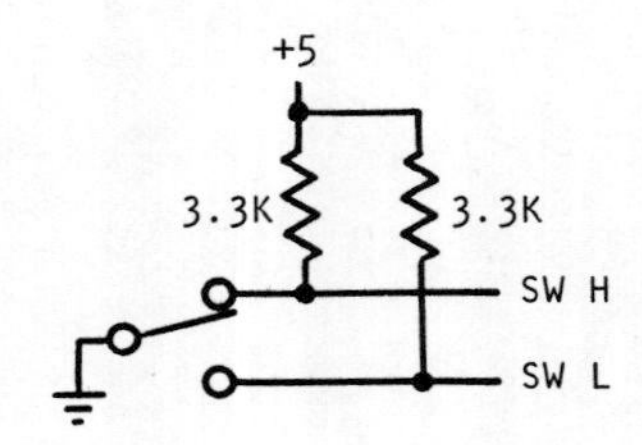

FIGURE 1-6 Switch circuit.

ground connected, the input structures can be damaged. Also, parasitic SCRs can be triggered so that if power is applied without removing the drive signal, a short-circuit condition will exist between the device's power and ground terminals.

SWITCHES

Toggle switches, slide switches, and push buttons are typical input devices found with logic breadboards. A single-pole, double-throw switch can produce two complementary logic outputs using the circuit of Figure 1-6. The 3.3K resistors provide sufficient drive in the 1 state for 20 TTL unit loads.

The contacts of a typical switch will bounce several times over a period of milliseconds after being operated. Thus, instead of the clean transition shown in Figure 1-7a, several pulses over a period of 10 to 50 milliseconds may be produced as shown in Figure 1-7b. However, the effects of contact bounce can be eliminated (the switch "debounced") using an S-R flip-flop (see Project A4). In a typical breadboard, toggle switches and slide switches are not debounced, while there may be built-in debouncing flip-flops for push buttons.

In the project descriptions in later chapters, we use "toggle switch" to indicate any switch which is not debounced, while "push button" indicates a debounced switch.

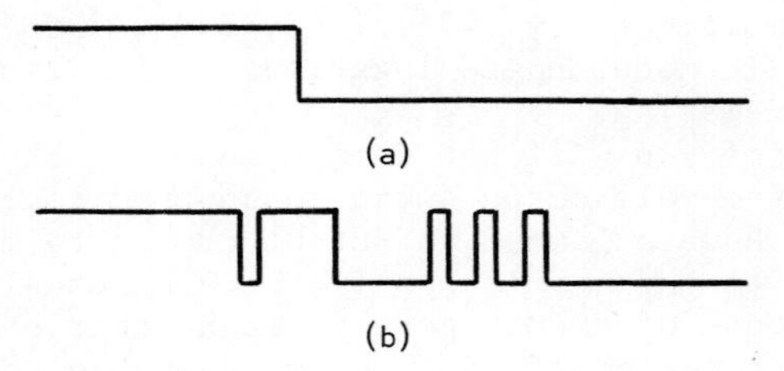

FIGURE 1-7 (a) Clean transition; (b) Contact bounce.

LAMPS AND SEVEN-SEGMENT DISPLAYS

Lamps and seven-segment displays are the principal output devices of logic breadboards. Lamp circuits usually contain an open-collector driver circuit and incandescent lamps. A hex open-collector inverter/driver such as a 7406 and a 4ESB lamp is a common combination. Each lamp is connected from an inverter output to +5 volts, so that a logical 1 on the inverter input lights the lamp, and a logical 0 extinguishes the lamp. The disadvantage of standard TTL drivers for lamp circuits in breadboards is that unused lamps will always be lit, since no-connection on a TTL input is interpreted as a 1. At Stanford we use SP391 drivers, which have PNP inputs. These drivers interpret a no-connection as a 0 input.

If your breadboard has seven-segment displays, it probably also has standard BCD drivers for the displays. A typical driver takes a 4-bit input and converts inputs 0000 through 1001 to the corresponding decimal digits on the display. Inputs 1010 through 1111 may cause either a blank or a strange character to be displayed. Hence, most seven-segment displays are useful for displaying information only in decimal or octal. However, some drivers, such as the Fairchild 9368, convert the codes 1010 through 1111 to the letters A, b, C, d, E, F, so that they can be used to display four bits in hexadecimal. Some breadboards also have BCD counters that driver seven-segment displays directly.

PULSE GENERATORS

Your breadboard may have a built-in pulse generator or other timing source. A pulse generator typically has two sets of controls -- one for varying the period (or frequency) of the output waveform and another for varying the length of the individual pulses. A single-step capability is also sometimes included, which allows you to single-step a circuit without disconnecting the circuit's clock input from the pulse generator and reconnecting it to a debounced push button.

Instead of a pulse generator, some breadboards have a fast crystal-controlled clock and a cascade of BCD counters to make lower frequency signals available. The breadboards used at Stanford have only a slow clock available. This is a 60 Hz square wave derived from the AC line by a simple analog circuit and Schmitt trigger.

TEST EQUIPMENT

In addition to breadboards, your laboratory will have a number of test instruments for observing and debugging the performance of your circuits. An oscilloscope is the most important test instrument in a digital logic lab. A 25 MHz, dual-trace scope is sufficient for debugging most microprocessor circuits and projects built with standard TTL.

A stand-alone pulse generator is necessary if your breadboard does not have one built in. Commercially available pulse generators typically provide variable pulse widths as short as 50 ns or less, and have variable frequencies up to 10 MHz or more. Some models have fixed TTL-level outputs, while others have outputs with variable voltage swings and polarity. If you use a pulse generator with variable output voltage, be sure to check for the proper swing (about 0 to 3.5 volts) before connecting it to a TTL circuit. An input voltage above 6 volts or below -1 volt will surely damage your circuit. Also, be careful not to connect pulse generator outputs incorrectly. Although most pulse generator outputs are protected against shorts to ground, a short to the +5 volt supply can damage the generator.

A hand-held logic probe is an extremely useful debugging device. A typical logic probe can be touched to a signal line to indicate whether it is at logic 0, logic 1, or an intermediate or floating level. Furthermore, some logic probes have built-in latches to catch pulses too short to give a normal visual indication. Project B4 discusses logic probes in more detail.

A companion device for a logic probe is a logic pulser. A logic pulser can emit a short (~100 ns), high-energy pulse that overrides any level being maintained by standard TTL outputs. For example, if a line is being held at logical 0 by a TTL output, at the user's command the pulser will pull the line up to logical 1 for 100 ns. A logic pulser is primarily useful for debugging printed circuit modules, since it allows test signals to be injected without breaking existing connections. For breadboard debugging, a pulser is less important since connections can be broken and test signals injected by hand fairly easily.

Logic analyzers are excellent test instruments for debugging large synchronous sequential circuits and microprocessor systems. A basic logic analyzer has a clock input and 8 to 16 data inputs. At each clock pulse, the data inputs are strobed and stored in a digital memory. The stored input words are then displayed on an an oscilloscope screen, either as binary digits or in a timing-diagram format. From 16 to 256 or more words of memory are provided, so that a long record of data input words can be observed. Triggering circuits are usually provided so that storage of data is not started until a selected input word is observed. The triggering point can be moved in time so that a number of words both before and after the triggering word can be displayed. Some logic analyzers also have the capability of triggering on asynchronous input pulses, independent of the clock input. This feature is especially useful for detecting and diagnosing system "glitches."

Digital multimeters are found in most labs, but they are not essential to a digital logic laboratory. DC voltage measurements can be made with an oscilloscope, and resistance measurements are not often needed. However, unless your scope has a current probe, a multimeter is needed to measure currents as in Project A1. In the absence of a scope, the voltage measurements that can be made by a multimeter are quite useful in the initial stages of circuit debugging (see Chapter 3).

2. Documentation Standards

This chapter gives documentation standards that should be used in preparing project reports. The amount of documentation required varies with project complexity, but there are generally five distinct items that should be included:

1) Block diagram

2) Logic diagram

3) State diagrams

4) Word description

5) Answers to questions

The block diagram gives an overview of the functional modules of the system, while the logic diagram gives a detailed logical description of the circuit that is sufficient for constructing the circuit. State diagrams should be included to aid in understanding any complex sequential control circuitry that is used. The word description, in conjunction with the previous three items, should concisely explain the operation of the system and point out the use of any "tricks" or assumptions that are not obvious. Finally, any questions asked in the project description should be answered.

The remainder of this chapter describes standards for drawing block diagrams and logic diagrams. Although there are recognized standards for logic symbols, the standards for interconnecting

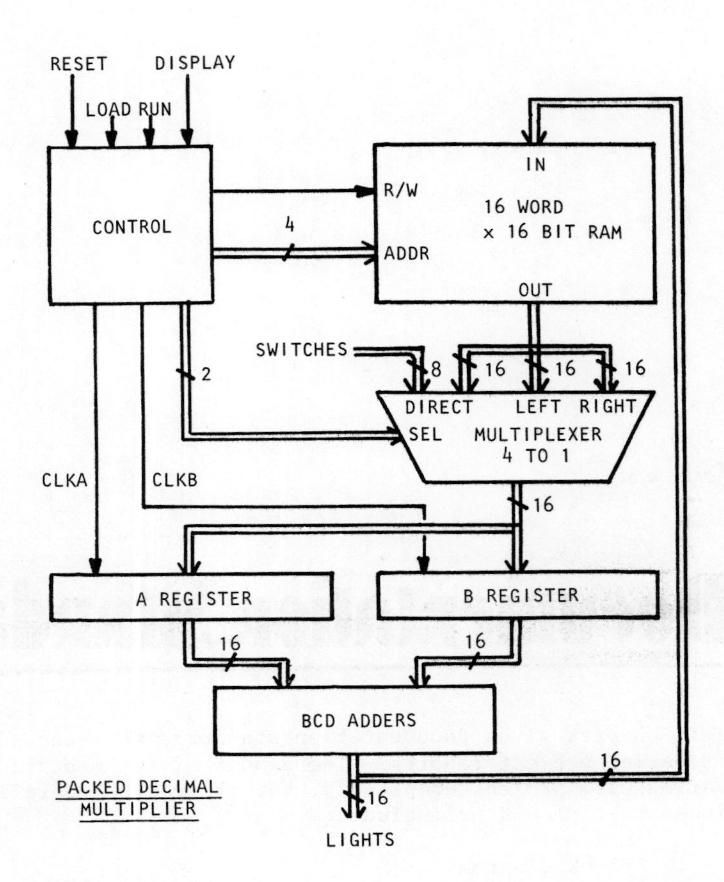

FIGURE 2-1 Block diagram.

symbols to form diagrams vary widely from company to company. We present standards that embody most of the important ideas of good drawing, and that lie somewhere between the most stringent and the most lax industrial standards.

BLOCK DIAGRAM STANDARDS

The purpose of a block diagram is to give a concise overview of the inputs, outputs, functional modules, data paths, and important control signals of a system. A block diagram must not be so detailed that it requires more than one page to draw, yet it must not be too vague. The block diagram for a small system (such as a B-series project) may have 3 to 6 blocks, while for a large system (D, E-series) may have 10 to 15 blocks (see Figures B4-1, C1-1, C2-1, C5-1, C6-1, C7-1, D1-3, D3-1, D4-1, D6-1, E3-1, E5-1). In any case, the block diagram must reflect what the designer considers to be the major subsystems, data paths, inputs, outputs, and control points of his

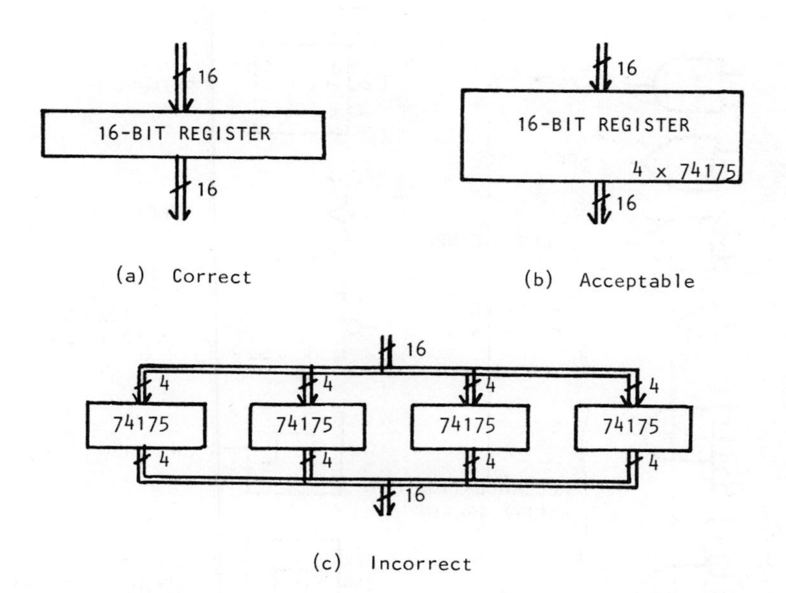

FIGURE 2-2 16-bit register block.

system. In very large systems additional block diagrams of individual subsystems may be needed, but there should always be one block diagram showing the entire system. (It is obvious that all diagrams must be drawn neatly using a straight edge for all lines.)

Figure 2-1 gives a sample block diagram. Notice that the diagram indicates the functional modules of the system, not the individual components that comprise it. For example, a 16-bit register to be implemented with four 74175's should be drawn as a single block (Figure 2-2a) rather than as four individual blocks (Figure 2-2c). However, it is acceptable and sometimes desirable to indicate on a block the specific components that will be used to implement it (Figure 2-2b).

As shown in Figure 2-1, buses are drawn with a double line. Data buses should have an indication of the number of lines in the bus. Specific control signals are drawn as single lines, while multiple control signals may be drawn as buses. The number of lines in a control bus may or may not be specified.

The flow of control and data in the block diagram should be clearly indicated. For logic diagrams we will require that signals generally flow from left to right, but this is not necessary in block diagrams. Inputs and outputs may be on any side of a block, and flow direction may be arbitrary. However, inasmuch as it improves readability, an attempt should be made to make the logic flow from left to right or top to bottom.

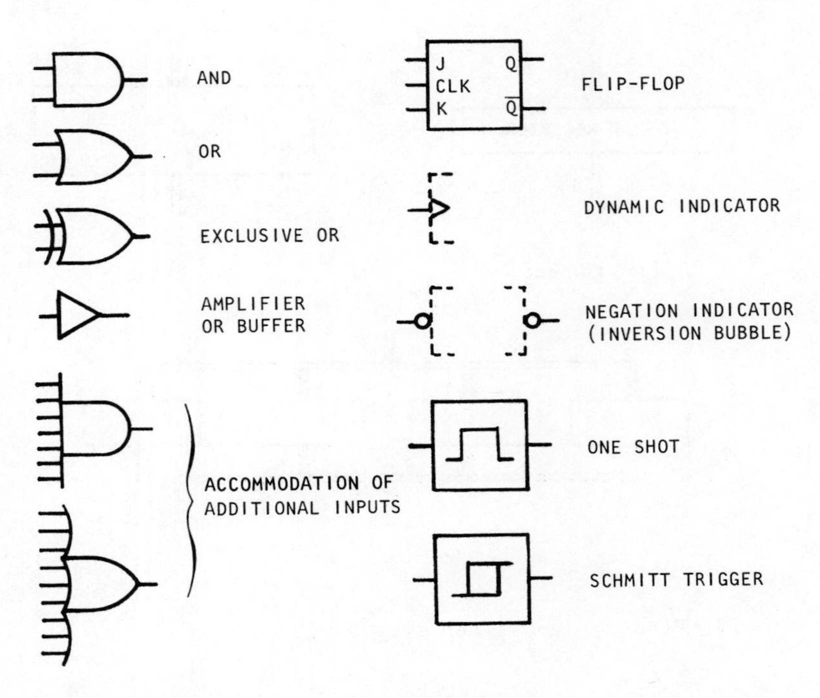

FIGURE 2-3 Logic standard symbols.

LOGIC SYMBOLS

Until 1973, there were two major standards for logic symbols: *American Standard Graphic Symbols for Logic Diagrams,* IEEE Std 91-1962, ASA Y32.14-1962, and *Military Standard Graphic Symbols for Logic Diagrams,* MIL-STD-806B, 25 February 1962. These standards have been superseded by a single standard, *IEEE Standard Graphic Symbols for Logic Diagrams,* IEEE Std 91-1973, ANSI Y32.14-1973. The revised standard allows both rectangular-shape and distinctive-shape symbols for logic elements; in this manual we will use the distinctive-shape symbols shown in Figure 2-3.

The symbols for NAND, NOR, EQUIVALENCE, and inverter gates can be obtained from the symbols for AND, OR, EXCLUSIVE OR, and buffer gates using the inversion bubble, as shown in Figure 2-4a. Equivalent symbols of Figure 2-4a and their equivalents in Figure 2-4b may be found in the same logic drawing; the rule for choosing which symbol to use in a particular case is given later.

Schmitt triggers and one shots may use the standard symbols shown in Figure 2-3. Schmitt triggers are also sometimes drawn as standard gate symbols with the hysterisis symbol inside. One shots with input logic show the input gating outside the rectangle. Open-collector outputs should be labeled with a "*" or "O.C."

The symbols for various flip-flops are given in Project A5. As an example, the symbol for a negative-edge-triggered J-K flip-flop is

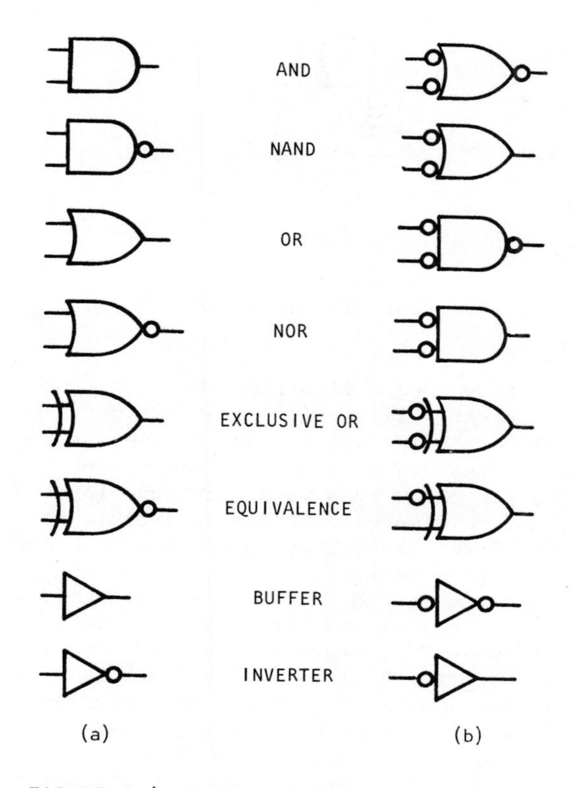

FIGURE 2-4 Logic symbols.

given in Figure 2-5. Note that inputs, except for preset (PR) and clear (CLR), are on the left and outputs are on the right. Note also that inputs are drawn in the same half of the symbol (top or bottom) as the outputs they control (J and PR with Q, K and CLR with $\overline{Q}$). Our flip-flop symbols do not correspond exactly to the IEEE standard, and the flip-flop symbols of various manufacturers differ widely.

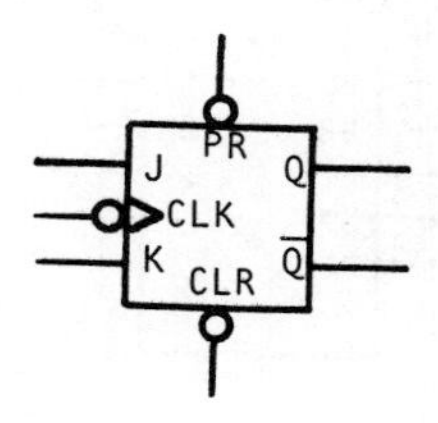

FIGURE 2-5 Negative-edge-triggered J-K flip-flop.

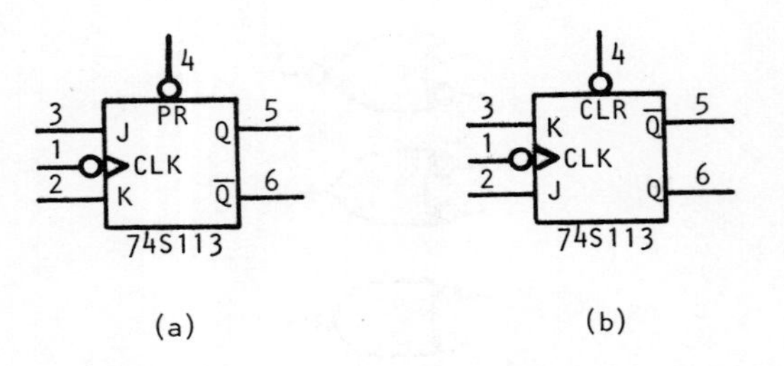

FIGURE 2-6 J-K flip-flop without clear.

Figure 2-6a shows a J-K flip-flop that does not have a clear input, only a preset. If a clear input but not a preset were needed, the symmetry of the flip-flop's J and K inputs allows the inputs to be conceptually renamed, as shown in Figure 2-6b. However the flip-flop should *not* be drawn this way. Although logic designers, technicians, and printed circuit layout artists all understand the equivalences of Figure 2-4, they do not always understand the equivalence of Figure 2-6.

MSI and LSI parts should be drawn as rectangles with labels inside to denote the part number and the names of input and output pins. In general, inputs should be drawn on the left and outputs on the right. The exceptions are cascading and other control leads, such as carries on an adder, clock inputs and outputs on a counter, and chip enable inputs, which may be drawn with outputs on the top and inputs on the bottom for clarity. For example, Figure 2-7a shows the symbol for a 2-input, 4-bit multiplexer and Figure 2-7b shows the symbol for a 4-bit adder. All occurrences of a particular MSI or LSI part should be drawn in the same format. Unused inputs and outputs should be drawn and indicated with an "NC" (no connection).

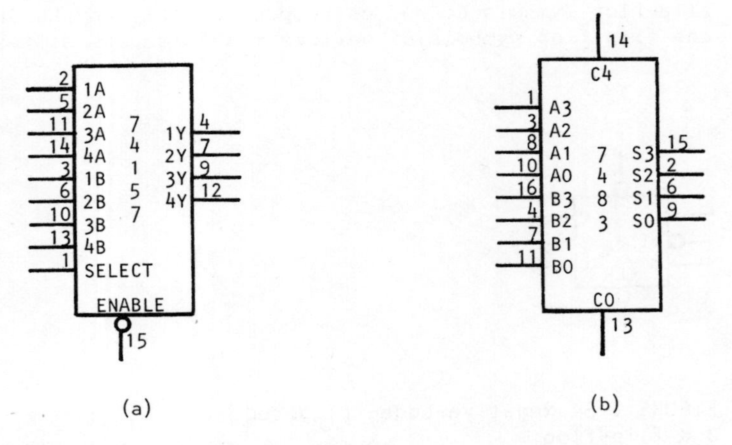

FIGURE 2-7 MSI parts.

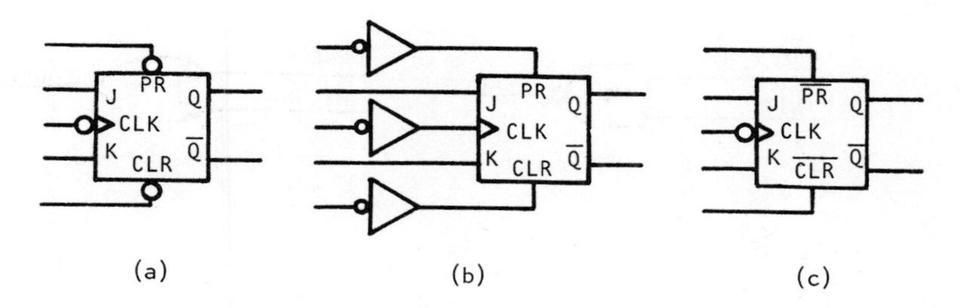

FIGURE 2-8 Interpretation of inversion bubbles.

Although it is generally required in industrial drawings, it is not necessary in your project drawings to draw the part numbers of standard SSI gates and flip-flops such as 7400, 7402, 7404, 7408, 7410, 7420, 7430, 7473, 7476, and 7486. Only special gates such as 7407, 7414, 7437, and so on need labels. However, the lines of each SSI, MSI, and LSI part should be labeled with a pin number, and each logic symbol should be labeled with a physical location, to facilitate wiring and debugging (for example, location B4 might indicate the fourth IC in the second column of your breadboard).

LOGIC CONVENTIONS AND INVERSION BUBBLES

Two conventions exist for the assignment of logical 0 and 1 signals to the voltages levels in a digital circuit -- *positive logic* and *negative logic*. Positive logic assigns the logic 1 designation to the more positive voltage level and logic 0 to the more negative; negative logic does the opposite. *We will use the positive logic throughout this manual.* Therefore, the designations 1 and HIGH are equivalent, as are the designations 0 and LOW.

Some designers think of inversion bubbles on gates, flip-flops, and MSI parts as inverting the logic convention from positive logic to negative or vice versa, but this only leads to confusion. Rather, an inversion bubble should be viewed simply as an abbreviation for an actual logical inverter. Thus the circuits of Figures 2-8a and 2-8b are equivalent. Notice also in this example that the label within a flip-flop or MSI rectangle refers to the function performed inside the rectangle, not at the input of the bubble if present. In Figures 2-8a and 2-8b, a logic 0 level on the $\overline{PR}$ line is required to preset the flip-flop. The flip-flop could also be drawn in the equivalent manner of Figure 2-8c, in which the inverted sense of the preset and clear functions is taken into account within the rectangle in the name of the terminal, and no inversion bubble is required. Figure 2-8a is the correct way of drawing the flip-flop symbol because it is the clearest.

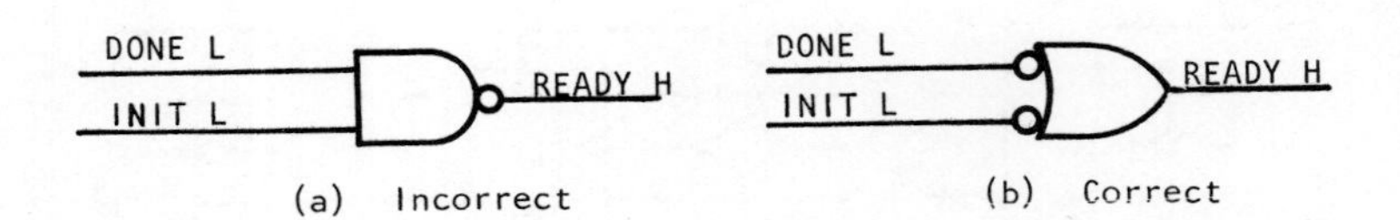

(a) Incorrect (b) Correct

FIGURE 2-9 Active levels.

SIGNAL NAMES, ACTIVE LEVELS, AND GATE SYMBOLS

Every signal line in a net work should have a descriptive label consisting of about 3 to 15 characters. The labels may contain spaces or underbars and perhaps some special symbols as well as alphanumerics, but only characters that are computer readable and printable should be used. Also no subscripts or overbars should be used. The foregoing requirements are necessary for most computer-aided design systems. Although some systems will generate a label for a line if none is supplied by the user, user-generated names are much more informative.

Each signal line should also have an active level associated with it. A signal is *active high* (labeled H or +) if it performs some action or denotes a named condition when it is high or on a low-to-high transition. A signal is *active low* (labeled L or -) if it performs some action when it is low or on a high-to-low transition. A signal is said to be *asserted* when it is at its inactive level. A signal is said to be *de-asserted* when it is at its inactive level. The active level of each signal in a circuit should be indicated at the end of its label. Since we are using positive logic, LOW and HIGH are synonymous with 0 and 1. The assignment of active levels to individual signals should not be confused with the logic convention (positive or negative), which is usually fixed for the entire circuit.

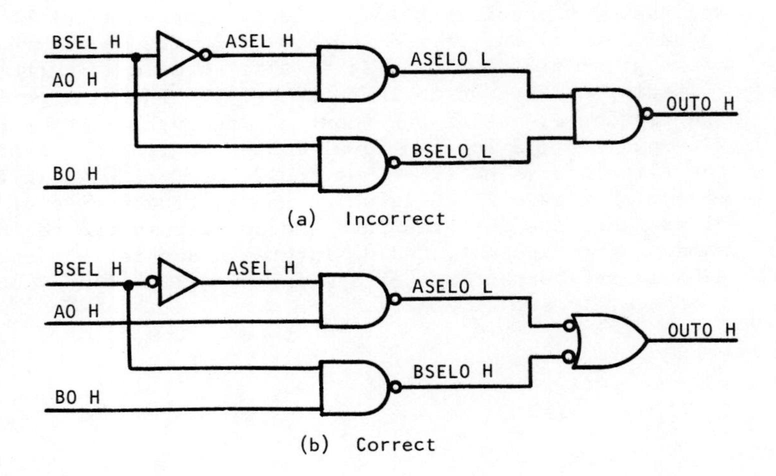

(b) Correct

FIGURE 2-10 Two-input multiplexer.

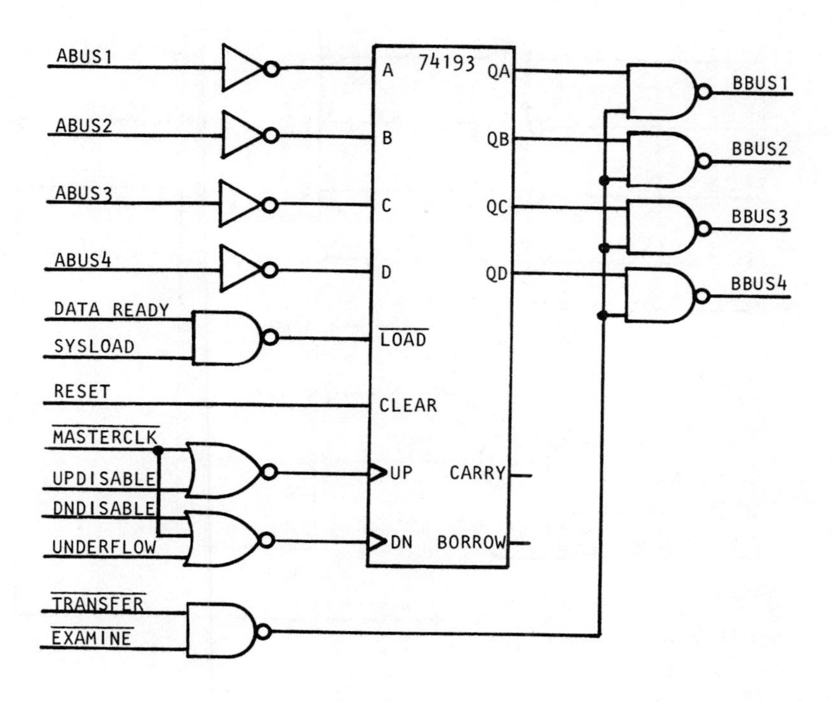

(a) Incorrect

FIGURE 2-11(a) Logic diagram standards example.

The alternative symbols for logic gates shown in Figure 2-4 should be used in a manner that is consistent with the active designations of signals. For example, a NAND gate used to produce an active-high signal when either of two active-low inputs is asserted should be drawn as shown in Figure 2-9b, not Figure 2-9a. Consider also the two ways of drawing one bit of a two-input multiplexer shown in Figure 2-10. The output NAND gate should be shown as an inverted-input OR, as shown in Figure 2-10b, indicating the designer's intention to OR together the two active-low signals ASELO L and BSELO L. Despite the fact that a simple, familiar two-level circuit like this is readily understood by most designers in either form, for unfamiliar and more complex circuits the use of proper gate symbols is a significant aid to understanding.

Consider the circuit shown in Figure 2-11a, which does not have active levels indicated and which does not use the alternative gate symbols where appropriate. When does the counter count up? When does the counter count down? When is the counter data transferred to the B bus? What is the relationship of the counter data polarity and count direction to the polarity of the data buses? These questions are much more easily answered when the circuit is drawn to standard, as shown in Figure 2-11b. It is apparent that both buses are active low, and the counter data is active high. The counter counts up when MASTERCLK and UPENABLE are asserted, and down when MASTERCLK and DNENABLE are asserted and UNDERFLOW is de-asserted. The counter is

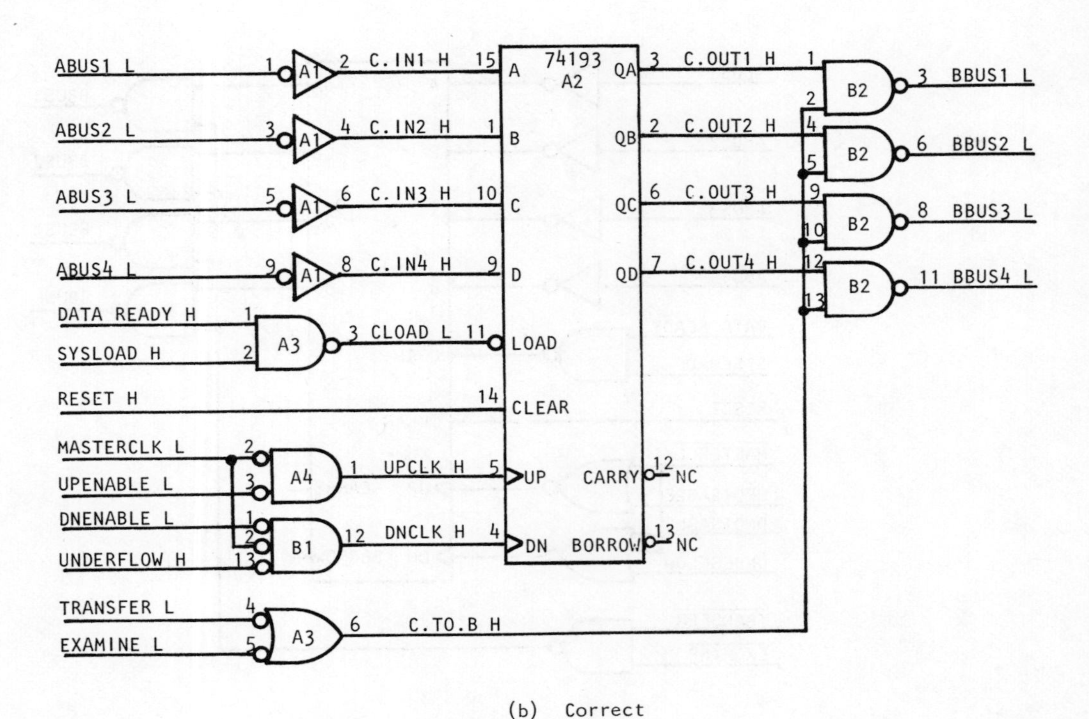

(b) Correct

FIGURE 2-11(b) Logic diagram standards example.

loaded when DATA READY and SYSLOAD are asserted, and the counter data is transferred to the B bus whenever TRANSFER OR EXAMINE is asserted. Figure 2-11b also shows pin numbers and IC locations.

The choice of names for signals in a circuit is up to the designer, but the choice of names affects the active level designations. Consider for example the circuit fragment of Figure 2-12. Both the 7495 and 74195 shift registers have an input that determines whether the register loads or shifts on the clock. The 7495 loads when this input is 1 and the 74195 shifts when this input is 1. In a circuit

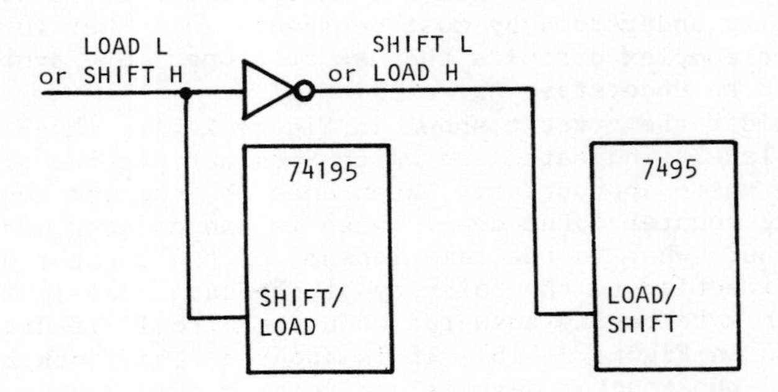

FIGURE 2-12 An example with many possible labelings.

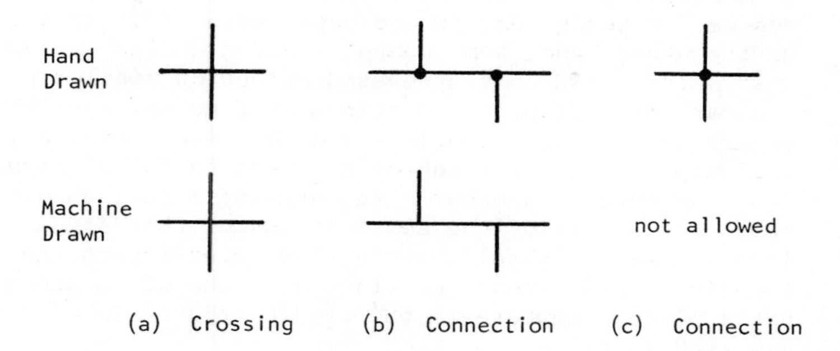

FIGURE 2-13 Line crossings and connections.

that uses both a 7495 and a 74195 connected as shown, two opposite names are possible for each of the two signals shown, for four possible total labelings.

DRAWING LAYOUT

Logic diagrams should be drawn with inputs on the left and outputs on the right for all parts, except those special cases mentioned earlier. System inputs should be on the left and outputs on the right, and the general flow of signals should be from left to right. If an input or output appears in the middle of the page, it should be extended to the left or right edge, respectively. In this way, a reader can find all inputs and outputs by looking at the edges of the page only. The most significant bits should be drawn at the top and the least significant bits at the bottom of the page. The lines of buses must be shown individually. All paths on the page should be connected when possible; paths may be broken if the drawing gets crowded, but the breaks must be flagged in both directions (see below).

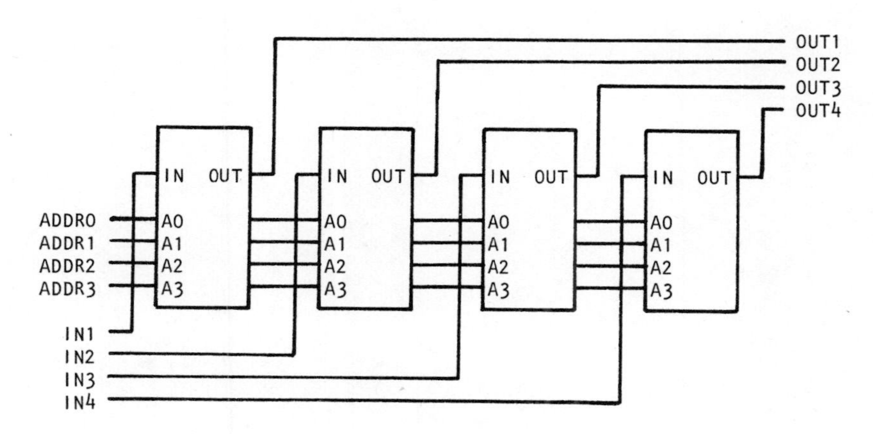

FIGURE 2-14 Drawing arrays.

Most designers feel that it is best to fit the logic diagram for a system (or subsystem) on one page, even if it is a fairly large page. On the other hand, some computer-aided design systems (and some instructors) can only process drawings on smaller pages, such as a standard $8\frac{1}{2} \times 11$ page. Drawings that occupy more than one page, regardless of size, should use a coordinate system (like that of a road map) to flag the source and destination of signals that travel from one page to another. An outgoing signal should have flags referring to all of the destinations of that signal, while an incoming signal should have a flag referring to the source only. An incoming signal should be flagged to the place where it is generated, not a place somewhere in the middle of a chain of destinations that use the signal.

It is easy to draw block diagrams without crossing lines to give a neater appearance (see Figure 2-1), but for logic diagrams this is not recommended. For logic diagrams it is best to allow lines to cross and to indicate connections clearly with a dot. Some computer-aided design systems do not have facilities for making connection dots in logic diagrams. To allow distinction between crossing lines and connected lines they adopt the convention that only "T" type connections are allowed, as shown in Figure 2-13.

One helpful shortcut in drawing the logic diagrams for large systems is shown in Figure 2-14. When drawing large arrays of MSI and LSI parts (such as memories) that all receive common inputs, the common lines may be run "under" instead of around the individual rectangles.

3. Design and Debugging Techniques

This chapter describes some techniques to apply and pitfalls to avoid for a better logic design. It also tells you what to do after you have designed and built your circuit following the guidelines and you discover that it still does not work.

DESIGN

The guidelines in this section will help you create a more reliable, maintainable, and debuggable circuit design. A good reference for efficient design with MSI and LSI in general is Blakeslee [1975].

The most important prerequisite for good design and debugging is good documentation. You should design your system completely *before* you build it, not while you build it, and documentation should be ready when you start building. Your assembly drawings may be rough, but they should be complete circuit diagrams for your system, giving part numbers, pin numbers, and physical locations for all components. After the system is built and debugged, a neat final schematic should be drawn according to the guidelines of the previous chapter.

The first design principle starts with initialization of your system. Circuit states should be reset on level signals, not on edges. Once the system is running, idle states in control circuits should be reset periodically, not just once at system initialization time (see Figure 3-1).

Hangup states with no exit should not exist. All invalid states should go to valid ones during periodic resets. Even if circuits were perfectly reliable so that an invalid state was never entered because of a fault, invalid states would still be entered during

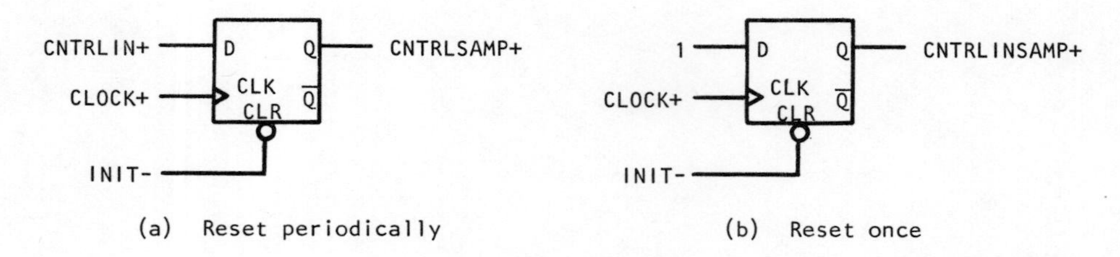

FIGURE 3-1 Periodic resets.

debugging and maintenance of the system.

Static and dynamic hazards can occur in many places. Their existence can be predicted by classical techniques, for example see McCluskey [1965]. However, there are many common hazards for which no theory is needed. The most common occur in decoding the states of a control counter. For example, the NAND gate that detects the 1111 state of the counter in Figure 3-2 will certainly produce "glitches" if it is a ripple counter such as 7493, and it can produce glitches due to unequal flip-flop delays even if the counter is synchronous, such as a 74191. Therefore such signals must not drive any asynchronous inputs such as clock or reset inputs of a flip-flop. Lines that may have glitches present must be sampled at a time when it is known that the lines are stable. If it is necessary to produce glitch-free outputs for control, then a Gray-code counter or shift register counter should be used (see Project A6). A typical system does not have a minimum state control machine, but instead has one flip-flop per conceptual control state or group of control states. For example, a CPU usually uses a RUN flip-flop rather than decoding the RUN state from an n-bit major state machine.

Races in sequential circuits can also be found by classical techniques [McCluskey, 1965], but there is one obvious class of races that designers often introduce. This occurs whenever both the synchronous and asynchronous inputs of a circuit depend on a common signal, and a change in that signal is used to activate the asynchronous inputs at the same time that it is causing a change in the synchronous inputs. An example of this problem, shown in Figure 3-3a, occurs in Project C1. Data must be loaded into a

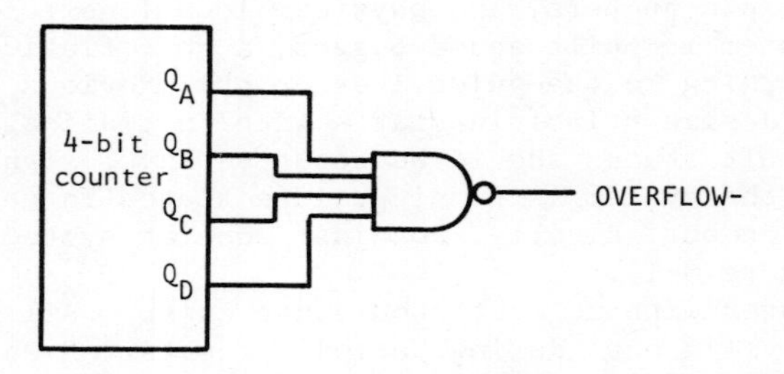

FIGURE 3-2 Decoding glitches.

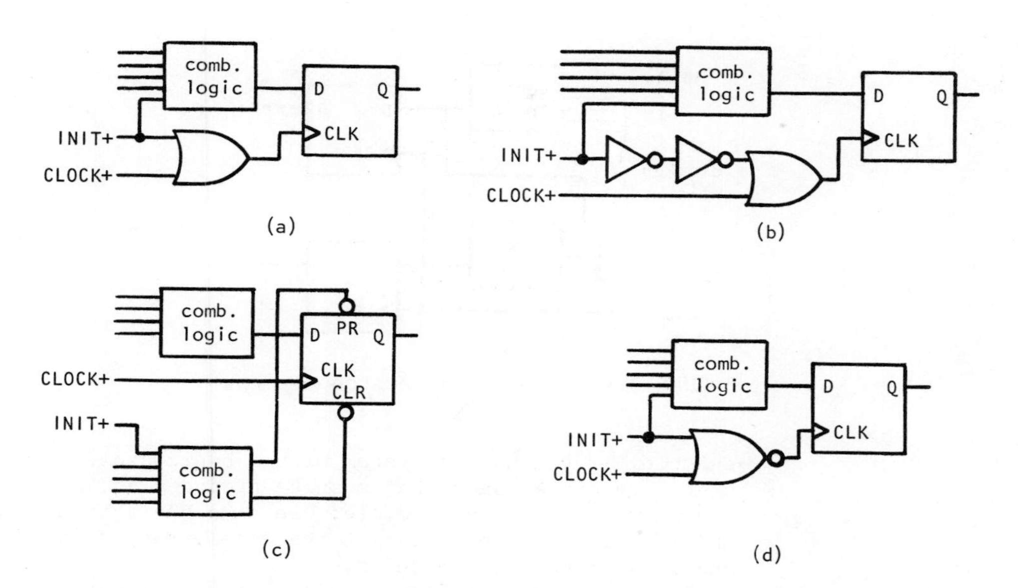

FIGURE 3-3 Simultaneous data and clock changes.

flip-flop both at system reset and during normal operation of the system. Therefore the flip-flop must be clocked by two signals, the normal clock (CLOCK+) and the reset signal (INIT+). However, the reset signal must also influence some combinational logic to set up the proper initial value in the flip-flop. Hence the reset signal causes almost simultaneous changes at the clock and data inputs of the flip-flop, and the designer hopes that the data will change before the clock. Since the combinational logic delay for the data is much longer than for the clock, designers sometimes compensate as shown in Figure 3-3b, a poor practice. A better solution is to use the preset and clear inputs of the flip-flop for resetting, as shown in Figure 3-3c. If the preset and clear inputs are not available, as in a 4-bit register, an acceptable solution is shown in Figure 3-3d. This solution can be used only if the analysis shows that the combinational logic delay for the data is clearly longer than for the clock. Then the combinational logic can be set up while INIT+ is high, and the flip-flop can be clocked just as INIT+ goes low.

Whenever possible, systems should be designed with single-step capability. That is, it should be possible to operate the system at any desired speed less than the maximum. This is extremely important for easy debugging. Single-step capability is not difficult if you use all clocked logic with no pulse generating or other asynchronous logic. All of the projects in this manual can be designed this way unless dynamic MOS RAMs or shift registers are used.

In general, you should avoid one shots and other pulse generators such as inverter trains in your designs. Resistors and capacitors for delay should also be avoided -- they are not reliable. Commercially available tapped delay lines are more accurate and reliable sources of short delays, but even these should be avoided if possible. For

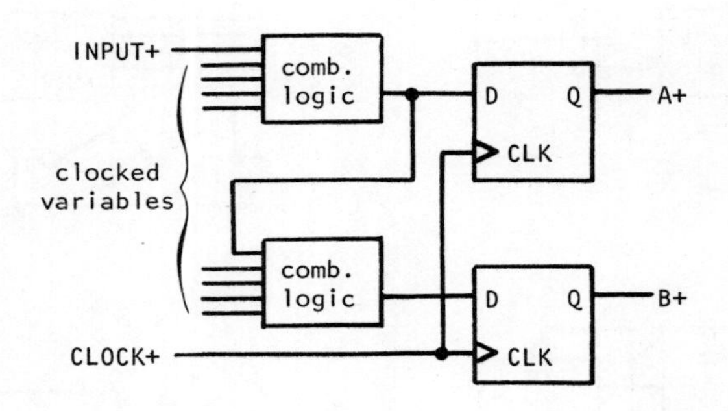

FIGURE 3-4 Sampling asynchronous inputs.

example, if you have a system with a basic cycle time of 200 ns, but there are clock edges that must be produced at 40 ns, 120 ns, and 160 ns within the basic cycle, use a 25 MHz system clock and a 5-bit shift register counter to get the edges, rather than a 5 MHz clock and delay lines or one shots.

When asynchronous input variables are present, such as in interfaces, flip-flops are used to sample the input variables at clock transition times. However, if this is not done properly, problems can occur. For example, in Figure 3-4 the asynchronous input, INPUT+, affects both flip-flop outputs A+ and B+. If INPUT+ changes just before a clock transition, the change may be picked up by flip-flop A but not by flip-flop B. The solution here is to sample INPUT+ with a flip-flop before the combinational logic so that inputs to flip-flops A and B do not change before the clock transition.

Even if flip-flops are used for synchronization, strange problems can sometimes occur. If an asynchronous input change occurs within a small window of time around the clock change, there is a small probability that the flip-flop will go into oscillation or into a metastable state halfway between logic 0 and logic 1. Furthermore, the length of time that the flip-flop may remain in the metastable state is theoretically unbounded. More details on this behavior can be found in an article by Chaney and Molnar [1973]. Since these events occur rarely and since the length of time of the anomalous behavior is usually short (typically less than 20 ns), the simplest solution to the problem is to cascade synchronizers. In this way the anomalous behavior of the primary synchronizer is eliminated by sampling its output once clock cycle later.

DEBUGGING

A small circuit, say less than five ICs, will sometimes work properly the first time it is turned on. All other circuits require debugging. The power and ground connections of all circuits should be checked before power is first applied. Large circuits, say over 30 ICs, should have all their wiring checked before applying power. Wiring can be checked by using an ohmmeter to perform a continuity check. Whenever it is found that a connection is missing, before making the connection one should see if there is simply a misplaced connection by checking all of the other wires leaving the two nodes involved.

After checking for smoke when the circuit is first powered up, an optimist will try to exercise the most complicated features of the system. However, since the system usually does not work, it is necessary to start debugging at a lower level -- Is the system state being reset properly? Are the proper clock signals reaching all of the sequential circuits? Are the data paths connected properly, and so on?

Two types of errors will be found in debugging: wiring errors and design errors. Theoretically all wiring errors can be detected by careful double-checking. However, if the system is almost working then it may be faster to try to deduce the location of wiring errors rather than to trace all of the connections. Both wiring and design errors are detected by working backwards in the circuit from some point that has a predictable behavior. If a signal at that point does not have the expected behavior, then work back *one level*. Do not try to guess the trouble source if this requires working back several levels in the logic. Such a procedure is usually just confusing since the problem is never where you expect it to be.

The first check to make on the powered-up circuit is a static check. You will have to get the circuit into a known state and then check to see that all signals are at their proper logic values. The next check is to single-step the system and check that the proper transitions are made. A logic probe that has the ability to catch pulses is extremely useful at this stage -- if sequential logic is being clocked by glitches, the logic probe can be used to trace their source (see Project B4). Finally you can debug the system at full speed.

The most common wiring errors are omitted and misplaced wires. An omitted wire usually results in an unconnected or "floating" input to a device. Floating inputs can be detected with a three-state logic probe (see Project B4) or with a scope or voltmeter. A good TTL logic 0 level is between 0.2 and 0.4 volts, and a logic 1 level is typically between 2.4 and 3.5 volts. A floating TTL input will show a voltage of about 1.6 to 1.8 volts. Logic levels for CMOS are equal to the supply voltage and ground, for all practical purposes. Misplaced wires sometimes result in short circuits between device outputs. Two shorted TTL outputs trying to maintain opposite logic levels will produce an output voltage of about 0.6 volts.

The most common design errors involve the disposition of unused inputs on MSI parts. Quite often when an input is not used, the designer forgets to make any connection to it, resulting in an effective logic 1 in TTL circuits. In CMOS circuits, because of the very high input impedance, very little noise can cause the input to drift back and forth between an effective logic 1 and an effective logic 0. Therefore, all unused inputs *must* be tied to a logic source in CMOS, and it is good practice to tie all unused TTL inputs to a logic source, even if a logic 1 is desired. Such problems as a counter failing to count or a register failing to load can often be traced to unused inputs.

The next most common design errors involve the 1s and 0s catching behavior of master/slave J-K flip-flops. Most designers quickly learn that a J-K flip-flop can respond to signals that appear and disappear well before the triggering transition. However, they often overlook the fact that this behavior is also present in MSI parts that use J-K flip-flops internally. For example, the use of J-K flip-flops internally in 74190 and 74191 counters imposes the restriction that the enable and up/down inputs must not be changed before the triggering clock edge.

We conclude by again emphasizing the importance of proper design and documentation -- if a circuit is designed and documented well, it is possible to debug all but speed problems using only a voltmeter and single-stepping the clock.

4. A-Series Introductory One-Week Projects

A1 - TTL CHARACTERISTICS

SUMMARY

This project is intended to demonstrate the electrical characteristics of the basic TTL gate, and in doing so provide a better understanding of TTL logic levels and fan-out considerations.

BASIC TTL GATE

The basic two-input TTL NAND gate (1/4 of a 7400) is shown in Figure A1-1. The power supply V_{CC} is +5 volts. When all inputs are HIGH (> 2 volts) the current in R1 flows through the collector of Q1 into the base of Q2, turning on Q2. This turns on Q4 and turns off Q3, and the output voltage is LOW. If any input goes LOW (< 0.8 volts), the current in R1 flows through the emitter of Q1, out of the input lead into ground. Q2 is turned off, turning off Q4 and turning on Q3, resulting in a HIGH output voltage. Except during transitions, transistors Q1, Q2, and Q4 are always either saturated or cut off. The output structure consisting of Q3 and Q4 is called a *totem pole*. Q3 is sometimes replaced in this structure by a Darlington pair to increase the HIGH output drive capability (see References).

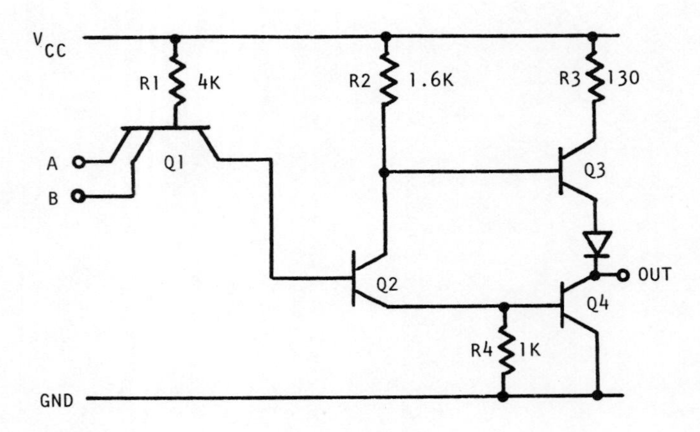

FIGURE A1-1 TTL NAND gate.

LOGIC LEVELS, NOISE MARGINS, AND FANOUT

The voltages corresponding to logic levels of standard TTL are depicted in Figure A1-2. V_{OL} is the guaranteed maximum output voltage in the LOW state, 0.4 volts for standard TTL. V_{OH}, the guaranteed minimum output voltage in the HIGH state, is 2.4 volts. V_{IL} is the maximum input voltage guaranteed to be recognized as a LOW. Since V_{IL} is 0.8 volts, it exceeds V_{OL} by 0.4 volts, and there is a *DC noise margin* of 0.4 volts in the LOW state. V_{IH}, the minimum input voltage guaranteed to be recognized as a HIGH, is 2.4 volts. Thus there is a DC noise margin of 0.4 volts in the HIGH state also.

All of the parameters above are guaranteed by the manufacturers of TTL over a specified range of temperature, power supply voltage, and

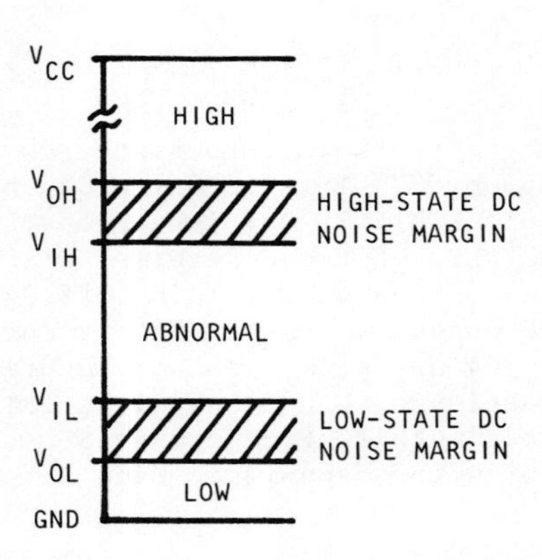

FIGURE A1-2 TTL logic levels.

fanout. The maximum fanout of standard TTL is 10, as we shall see below.

In the LOW state, a standard TTL input requires 1.6 mA of sink current (guaranteed maximum). A standard TTL output is guaranteed to sink at least 16 mA of current in the LOW state. The 1.6 mA is sometimes referred to as a *LOW-state unit load*.

In the HIGH state, a standard TTL input requires a source of 40 μA of current (guaranteed maximum). A standard TTL output is guaranteed to source at least 400 μA of current in the HIGH state. Hence one standard TTL output can drive 10 standard TTL inputs in the HIGH state. The 40 μA is sometimes referred to as a *HIGH-state unit load*.

TTL gates with Darlington pairs in the totem pole output have more driving capability in the HIGH state. They can typically source 800 μA and hence they have a fanout of 20 in the HIGH state only. What good is increased fanout in only the HIGH state? The answer is as follows. When two inputs of the same gate are tied together, as in a two-input NAND gate used as an inverter, the LOW-state input current is only 1.6 mA and the two tied inputs look like only one LOW-state unit load. However, in the HIGH state, the tied inputs look like two unit loads. Hence, the increased HIGH-state fanout capability allows gate inputs to be tied together without reducing the total number of gates that can be driven.

Other TTL families have different output voltage levels and input and output currents. These are described in a subsequent section. When dealing with a single family it is usual to define unit loads to facilitate fanout calculations. However, when different families are interconnected it is necessary to sum the individual input currents and compare with the output drive capability to check that fanout capabilities have not been exceeded.

The effect of loading an output with more than its rated fanout is to increase its LOW-state output voltage and decrease its HIGH-state output voltage. Because of TTL's DC noise margins, a slightly overloaded circuit will still work in noise-free conditions, but of course the noise margins are reduced.

Unused inputs of TTL gates left "floating" behave as if they have a logic 1 applied, but a small amount of noise can change this to a 0. Therefore, unused TTL inputs should be tied to a source of logic 1. Tying directly to the 5 volt supply is not recommended since a transient of over 5.5 volts as short as a few nanoseconds can damage the gate. Rather, tie unused inputs to 5 volts through a current-limiting resistor (1K is good for 50 inputs), or use the output of an unused NAND gate with one of its inputs grounded.

OPEN-COLLECTOR OUTPUTS AND WIRED LOGIC

Transistor Q3 in the totem-pole output of Figure A1-1 is said to provide *active pull-up*, since the transistor actively pulls up the output voltage on a LOW-to-HIGH transition. This transistor is eliminated in gates with *open-collector outputs*, as shown in Figure A1-3. An external resistor provides *passive pull-up* to the HIGH level. The main disadvantage of open-collector gates is that their LOW-to-HIGH output transitions are slower than those of gates with active pull-up. Open-collector outputs are used in three major applications: driving lamps and LEDs, performing wired logic, and busing.

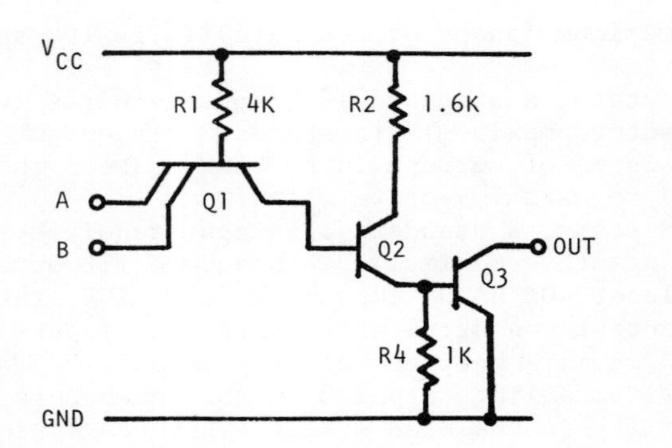

FIGURE A1-3 Open-collector TTL.

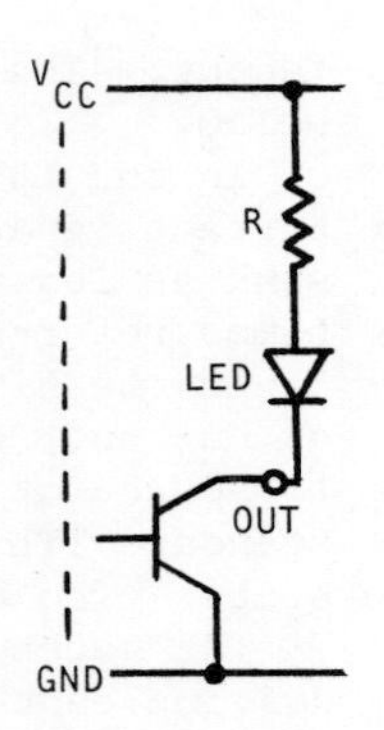

FIGURE A1-4 Driving an LED.

An open-collector output can drive an LED as shown in Figure A1-4. The value of the pull-up resistor R is chosen so that the proper current flows through the LED. For example, if the LED requires 15 mA for full brightness and produces a voltage drop of 1.6 volts, we choose R = (5 - 1.6 - 0.4)/0.15 = 200 ohms.

If the outputs of several open-collector gates are tied together with a single pull-up resistor then wired logic is performed. An AND function is obtained since the wired output is HIGH if and only if each individual gate output is HIGH. Any output going LOW is sufficient to pull the wired output LOW. Note that wired logic cannot be performed using gates with totem-pole outputs. Two totem poles wired together and trying to maintain opposite logic values produce a voltage in the range of 0.6 to 1.0 volt, not a normal logic voltage.

Open-collector gates can be tied together to allow several devices to share a common bus. At any time all but one of the gates on the bus are in their HIGH state. The remaining gate may stay in the HIGH state or drive the bus to the LOW state depending on whether it wants to transmit a logical 1 or 0 on the bus. Control circuitry is used to select the particular device that drives the bus at any given time.

A proper choice of value for the pull-up resistor must be made in open-collector applications. Two calculations are made to bracket the allowable values of R as follows.

MINIMUM - The sum of the current through R in the LOW state and the LOW-state input currents of gates driven by the wired outputs must not exceed the LOW-state output driving capability, 16 mA for standard TTL.

MAXIMUM - The voltage drop across R in the HIGH state must not reduce the output voltage below V_{OH}, the guaranteed minimum HIGH state output voltage of 2.4 volts. This drop is produced by the HIGH-state output leakage current of the wired outputs and the HIGH-state input currents of the driven gates.

For example, suppose four open-collector outputs are wired together and the wired output drives two standard TTL inputs. To stay within the LOW-state drive of 16 mA, the current through R may be no more than (16 - 3•1.6) = 11.2 mA. Hence the minimum value of R is (5 - 0.4)/.0112 = 411 ohms. In the HIGH state, typical open-collector gates have a leakage current of 250 μA. Hence the HIGH-state current requirement in the example circuit is (4•250) + 2•40) = 1080 μA. This current must produce a voltage drop of no more than (5 - 2.4) = 2.6 volts, and thus the maximum value of R is 2.6/.00108 = 2407 ohms. Hence any value of R between 411 and 2407 ohms can be used. A lower value is generally preferred, since it increases the HIGH-state noise margin and increases the speed of LOW-to-HIGH transitions.

THREE-STATE OUTPUTS

Although open-collector devices can be used for busing applications as described in the last section, the use of devices with *three-state outputs* has become much more popular. Three-state devices have an extra input, usually called "output enable" or "output disable," for placing the device's output(s) in a high impedance state. Thus the output can have three states -- logic 0, logic 1, or "Hi-Z."

A three-state bus is created by wiring several three-state outputs together. Control circuitry for the "output enables" must ensure that at most one output is not in its Hi-Z state at any time. The single device not in its Hi-Z state can transmit logical information on the bus.

Devices with three-state outputs are designed so that the output enable delay is somewhat longer than the output disable delay. Thus if a control circuit asserts one device's output enable input at the same time it de-asserts another's, it is guaranteed that the second device enters the Hi-Z state before the first device leaves it. If two devices try to maintain opposing logical states on the bus at the same time the situation is similar to normal TTL -- the outputs are not damaged but a non-logic voltage is produced on the bus. The main reason for avoiding this situation is that it generates large transient pulses which can cause system noise problems.

There is a leakage current of typically 40 μA associated with three-state outputs in their Hi-Z state. This current, as well as the input currents of receiving gates, must be taken into account when calculating the maximum number of devices that can be placed on a three-state bus. In the LOW state, an active output must be capable of sinking 40 μA of leakage current for every other three-state output on the bus as well as 1.6 mA for every standard TTL input on the bus. In the HIGH state, an active output must be capable of sourcing 40 μA of leakage current for every other three-state output on the bus as well as 40 μA for every standard TTL input on the bus. A typical three-state gate, such as 74126, sinks a standard 16 mA in the LOW state but sources 5.2 mA in the HIGH state. This allows as many as 129 three-state outputs (and one standard TTL input) to be bused together.

TABLE A1-1 TTL Families

Family	74	74L	74H	74S	74LS
Power consumption per gate (mw)	10	1	22	20	2
Typical propagation delay (ns)	9	33	6	3	9
Speed-power product (pJ)	90	33	132	60	18
V_{IL} LOW-level input voltage (volts)	0.8	0.7	0.8	0.8	0.8
V_{OL} LOW-level input voltage (volts)	0.4	0.4	0.4	0.5	0.5
V_{IH} HIGH-level output voltage (volts)	2	2	2	2	2
V_{OH} HIGH-level output voltage (volts)	2.4	2.4	2.4	2.7	2.7
I_{IL} LOW-level input current (mA)	-1.6	-0.18	-2	-2	-0.36
I_{OL} LOW-level output current (mA)	16	3.6	20	20	8
I_{IH} HIGH-level input current (μA)	40	10	50	50	20
I_{OH} HIGH-level output current (μA)	-400	-200	-500	-1000	-400

OTHER TTL FAMILIES

The basic TTL gate of Figure A1-1 can be designed with different resistor values to produce gates with lower power consumption and slower speed, or higher speed and higher power consumption. The two resulting families are called *low-power TTL* (74L series) and *high-speed TTL* (74H series).

It was mentioned in the first section that transistors Q1, Q2, and Q4 in Figure A1-1 saturate in normal operation. The time needed to come out of saturation accounts for much of the switching time of standard TTL. Saturation can be eliminated by placing a Schottky diode between the base and collector of each saturating transistor. The resulting transistors, which do not saturate, are called *Schottky clamped transistors* or *Schottky transistors* for short. There are two Schottky families, differing mainly in resistor values. *Standard*

Schottky TTL (74S series) has the lower resistor values and the highest speed of all TTL families. *Low-power Schottky TTL* (74LS series) has higher resistor values and moderate speed.

Important characteristics of the five TTL families are given in Table A1-1. Low-power Schottky has become very popular in new designs, since it has the same speed as standard TTL but one-fifth the power consumption. TTL circuits designed for microprocessor interfacing usually have low-power Schottky inputs since TTL-compatible MOS microprocessor outputs usually cannot sink more than 2 mA in the LOW state. The main disadvantage of low-power Schottky is its reduced LOW-state DC noise margin.

Some TTL-compatible circuits for busing applications have *PNP input transistors* instead of the usual multiple-emitter transistor shown in Figure A1-1. This input structure has the advantage that it requires very little current in the LOW state as well as the HIGH state, typically 15 μA.

The TTL-compatible inputs of MOS microprocessors, memories, and other LSI circuits require only a few microamperes of current in either the LOW state or the HIGH state. Hence DC fanout of circuits driving these components is never a problem. More important is the capacitance associated with each input, on the order of 10 pf. A bus driver must be capable of charging or discharging the capacitance associated with the inputs it drives in a sufficiently short time. As the number of inputs on a bus increases, so does the switching time of 0-to-1 and 1-to-0 transitions. The fanout of MOS components like microprocessors is often specified as a combination of DC fanout plus capacitive drive. For example, a fanout of one standard TTL load plus 100 pf means that an output can drive ont TTL load without exceeding its DC current rating *plus* charge or discharge 100 pf without exceeding its rated propagation delay specification.

ASSIGNMENT

I. For each of the following three experiments, you are to set up the indicated circuit and plot the variable values choosing appropriate scales that display their gross behavior. Use the same TTL NAND gate (1/4 of 7400) for all three experiments.

1) Output voltage versus input voltage:

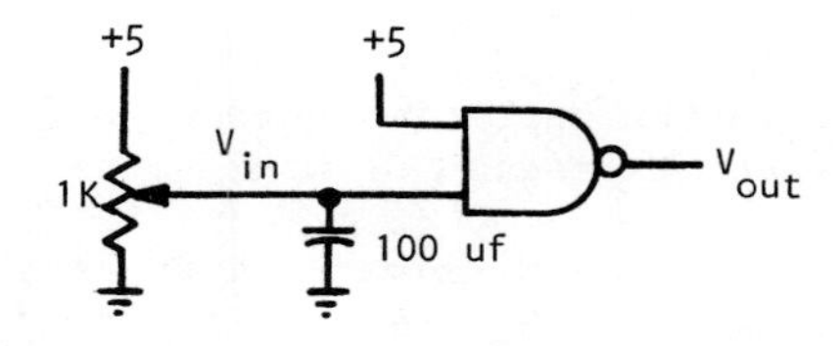

2) Input current versus input voltage:

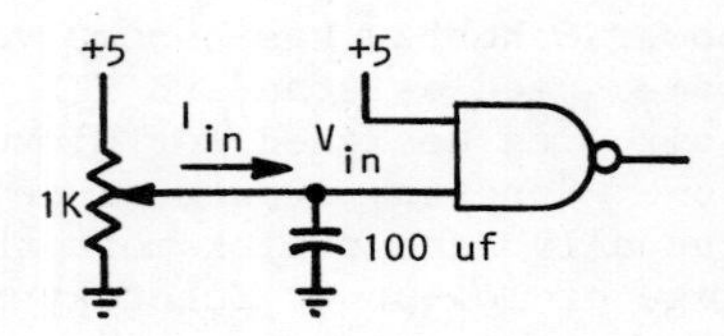

3) Output current versus output voltage:

a) Zero-state

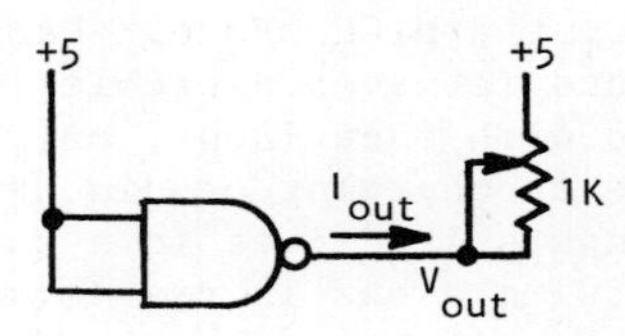

NOTE: In this part do not let I_{out} exceed 120 mA for more than a minute or the gate will be damaged.

b) One-state

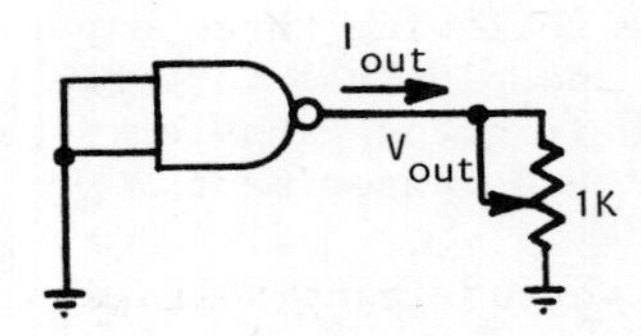

4) Questions:

a) What is the approximate voltage gain ($\Delta V_{out}/\Delta V_{in}$) of the gate at its region of maximum gain?

b) What is the approximate input impedance ($\Delta V_{in}/\Delta I_{in}$) of the gate at V_{in} = 0? Also at V_{in} = 2?

c) What would be the output voltage for a gate with no connection to the input (I_{in} = 0)? Would it be wise to use no-connection as a constant logic-value? Explain.

d) Using your results, calculate the maximum fan-out for an output in the zero state. You should assume a DC noise margin of 0.4 volt.

e) Again assuming a DC noise margin of 0.4 volt, what would the maximum fan-out be for an output in the one state?

f) Why do you think the manufacturer of the chip specifies 10 as the maximum fan-out?

II. Measure the following four currents:

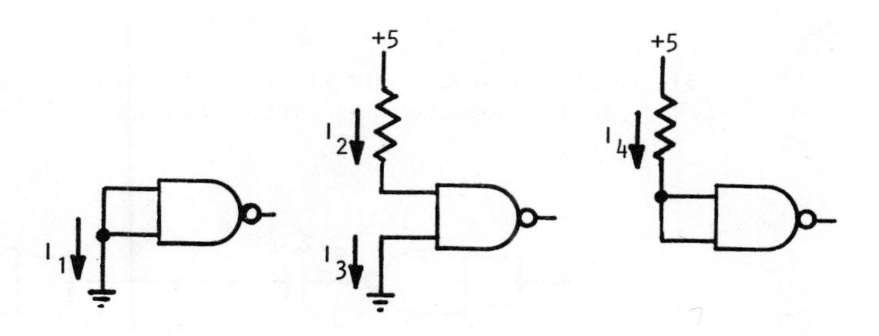

What does this imply about fan-out when several inputs of one gate are tied together?

III. Related problems:

1) The following circuit is intended to provide a logic value of one when the switch is open and zero when the switch is closed. Assuming 0.4 volt as the maximum zero-state input voltage, calculate the maximum value of R. How much power is wasted in the resistor when the switch is closed?

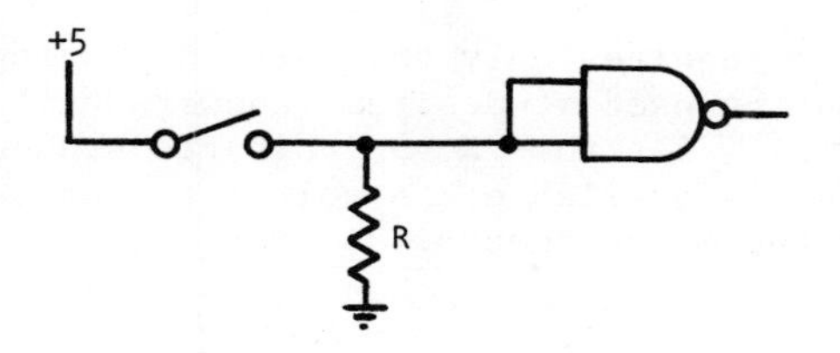

2) Do the same calculations for this circuit, assuming 2.4 volts is the minimum one-state input voltage.

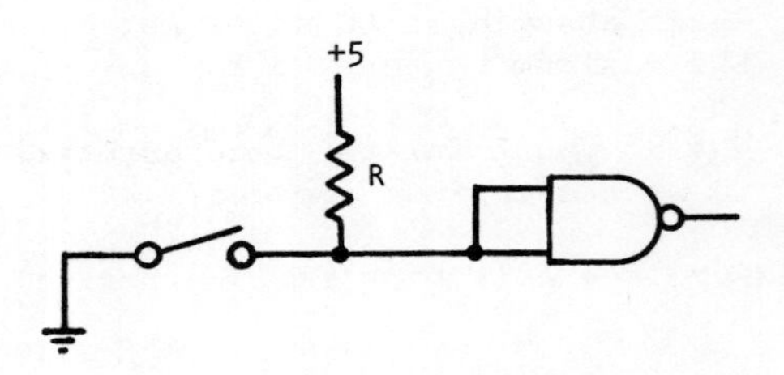

3) What would the output be for this circuit, with the indicated input? Assume the gates have zero delay.

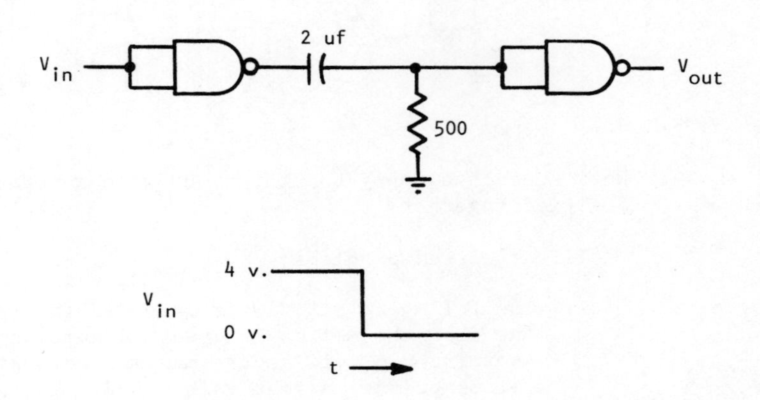

PARTS

This project requires only the parts appearing in the diagrams above. A 74LS00 may be used if desired (answers will be different from standard TTL). For the 1K variable resistance use a 10-turn potentiometer if one is available. A voltmeter and sensitive ammeter will be needed for the measurements.

REFERENCES

Descriptions of the electrical characteristics of TTL logic families can be found in several texts [Gschwind and McCluskey, 1975; Hnatek, 1973; Barna and Porat, 1973; Morris and Miller, 1971] and in manufacturers' catalogs and applications notes [Signetics, 1974; Texas Instruments, 1973; Fairchild, 1973, 1972]. Discussions of problems of noise and reflections in digital signal transmission may be found in Blakeslee [1975] and Morris and Miller [1971]. One of the best references on TTL characteristics and applications is Fairchild [1973]. Descriptions of a number of different logic families appear in Gschwind and McCluskey [1975], Hnatek [1973], Barna and Porat [1973], and Kohonen [1972].

A2 - CMOS CHARACTERISTICS

SUMMARY

In this lab we will investigate the DC and AC characteristics of a CMOS NAND gate using a +5 volt supply.

BASIC CMOS GATE

A basic two-input CMOS NOR gate is shown in Figure A2-1. Q1 and Q2 are p-channel MOS transistors, and Q3 and Q4 are n-channel MOS transistors. Most CMOS ICs will allow V_{CC} to be set at any value from +3 volts to +15 volts, depending on the application. V_{CC} and ground are labeled V_{DD} and V_{SS}, respectively, by some manufacturers.

The basic gate in Figure A2-1 performs the NOR function as follows. If both inputs are LOW then both Q1 and Q2 are on, and both Q3 and Q4 are off; the output is HIGH. If at least one input is HIGH, then at least one of Q1 and Q2 is off, and at least one of Q3 and Q4 is on; the output is LOW. A two-input CMOS NAND gate, shown in Figure A2-2, reverses the positions of series and parallel transistors to obtain the NAND function.

LOGIC LEVELS, NOISE MARGINS, AND FANOUT

The input characteristic of an MOS transistor is essentially capacitive,

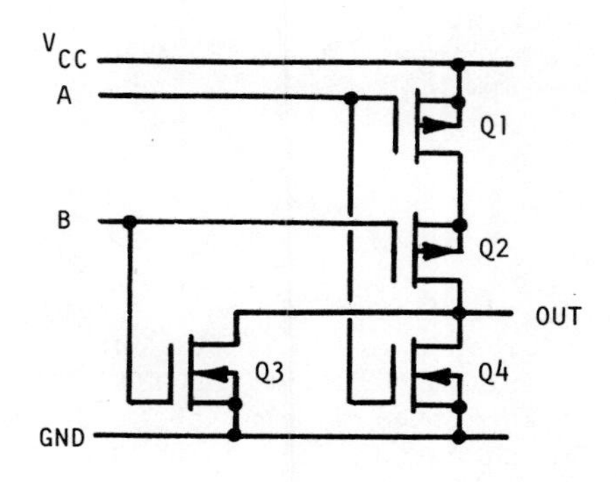

FIGURE A2-1 CMOS NOR gate.

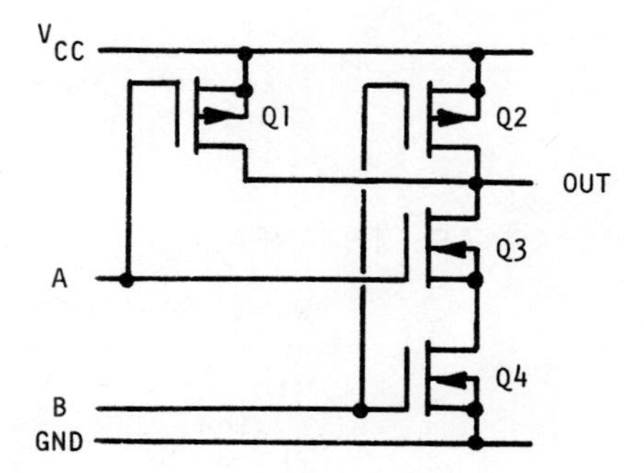

FIGURE A2-2 CMOS NAND gate.

looking like a 10^{12} ohm resistor shunted by a 5 pf capacitor. Thus the input impedance of a CMOS gate is very high. A CMOS output driving a CMOS input needs to supply almost no current, and hence the voltage drop across its active output transistor(s) is nearly zero. Therefore the logic levels seen in a CMOS system are essentially V_{CC} and ground.

CMOS circuits typically have a *noise immunity* of 0.45 V_{CC}. This means that an input which is 0.45 V_{CC} or less away from V_{CC} or ground will not propagate through the system as an erroneous logic level. This does not mean that the output will be corrected to the proper value of V_{CC} or ground, but it will be closer to correct value than the input was. After passing through a few gates, the error will be attenuated completely.

CMOS circuits also typically have a *DC noise margin* of 1 volt over the full power supply range. Stated verbally, the specification says that for the output of a circuit to be within 0.1 V_{CC} of the proper logic level (V_{CC} or ground), the input can be as much as 0.1 V_{CC} plus 1 volt away from the proper logic level. This is similar in nature to the standard TTL DC noise margin of 0.4 volts.

Since CMOS gates require almost no static input current, the DC fan-out of CMOS driving CMOS is virtually unlimited. However, current is required to charge and discharge the capacitance of CMOS inputs on logic transitions. The propagation delay of a CMOS gate is typically specified for a particular capacitive load, say 50 pf. If the capacitance of the load is higher, the propagation time is longer. As a rule of thumb a designer can assume the load will be 5 pf per CMOS input plus 5 pf to 15 pf for stray wiring capacitance.

TTL-TO-CMOS INTERFACE

When used with a 5 volt power supply, CMOS is somewhat compatible with TTL. In the LOW state, a TTL output can drive CMOS directly. However, the guaranteed TTL HIGH output level of 2.4 volts is not a valid input level for CMOS. If a TTL output drives only CMOS, then there is essentially no output current and the HIGH output level may be 3.5 volts or higher. Whether this is sufficient for a reliable interface depends on the exact manufacturer's specifications for both the TTL outputs and the CMOS inputs. A valid HIGH output level can always be ensured by tying a pull-up resistor from the TTL output to the 5 volt supply.

CMOS-TO-TTL INTERFACE

When CMOS drives TTL the HIGH state is no problem. The crucial question is whether CMOS can sink TTL input current in the LOW state without exceeding the maximum value of the TTL LOW-state input voltage. Typical CMOS gates are specified to sink about 0.4 mA in the LOW state while maintaining an output voltage of 0.4 volts or less. This is sufficient to drive two low-power TTL inputs or one low-power Schottky input, but it is unsufficient to drive standard TTL. Close examination of the specifications may show that it is possible to drive standard TTL at room temperature with some loss of DC noise margin. However, it is better to use a special buffer such as a 74C901 to drive standard TTL from CMOS.

ASSIGNMENT

I. DC Terminal Characteristics of a 74C00 NAND Gate

1) Graph the output voltage versus the input voltage, using a power supply for the input voltage generator, for the range $0 \leq V_{in} \leq 5$ volts.

2) Connect the output of the 74C00 gate to its own inputs. What voltage appears at the output? Do the same experiment on a 7400 TTL gate and compare results. Why does the 74C00 behave differently?

3) With both inputs at logical 1, attach the output of the 74C00 to several TTL gate inputs of the type 7400. Also connect one of the 7400 outputs to a lamp. How many TTL inputs may be connected before the lamp goes out? Plot the voltage at the 74C00 output as a function of the number of TTL inputs attached to it. How many standard TTL loads is the 74C00 rated to drive? How many CMOS loads can the 74C00 drive under your ideal lab conditions?

II. AC Terminal Characteristics of a 74C00 NAND Gate

1) Using a pulse generator as the input, measure the rise time, fall time, and propagation delay of the 74C00. Are the HIGH-to-LOW and the LOW-to-HIGH propagation delays the same? Explain. Are the rise and fall times the same? Explain.

2) Now attach a 100 pf capacitor to the output of the 74C00, with the other lead at ground. What are the rise and fall times now? If the input capacitance of a 74C00 gate is 5 pf, what maximum fan-out would *you* use in a CMOS system before deciding to buffer a signal to drive more gates?

3) Construct the circuit below:

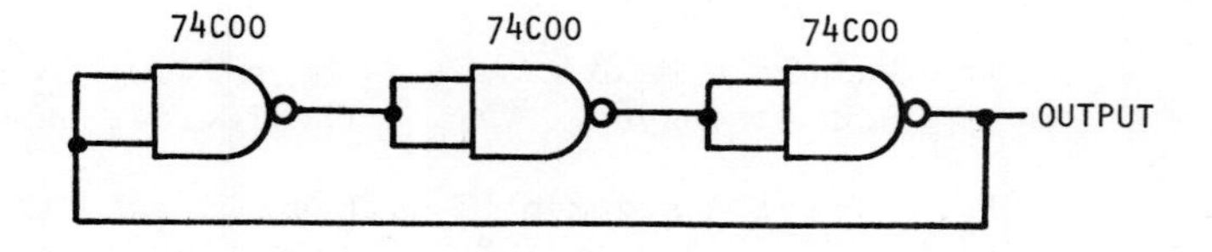

What waveform do you observe on the scope? Why is this different from the waveform in Assignment I(2)?

4) Construct the circuit below:

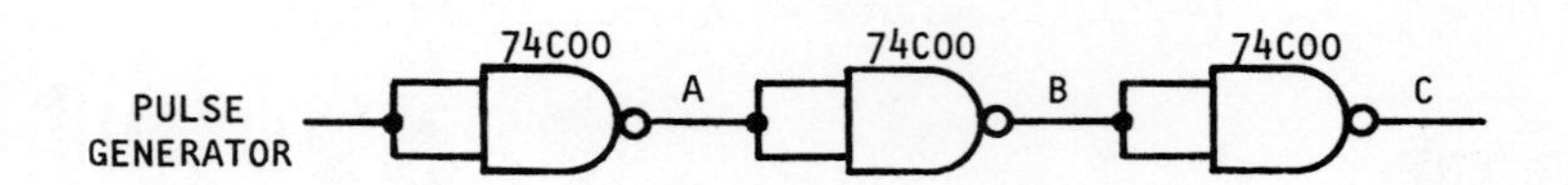

With the pulse generator set for a 0 to 1.9 volts square wave, sketch the waveforms at points A, B, and C. Repeat the above for 2, 2.1, and 2.2 volt square waves. What can you deduce about noise margins in CMOS? What are the advantages and disadvantages of CMOS versus TTL?

PARTS

This project requires one CMOS NAND gate such as 74C00, one TTL NAND gate such as 7400, and a 100 pf capacitor.

REFERENCES

Descriptions of the electrical characteristics of CMOS gates can be found in a number of texts [Gschwind and McCluskey, 1975; Hnatek, 1973; Barna and Porat, 1973] and in the manufacturers' catalogs and applications notes [National, 1974; Motorola, 1975]. The narrative material in this project is based on National Semiconductor application notes AN-77 and AN-90, which can be found in National [1974]. A good introduction to CMOS can be found in Melen and Garland [1975].

A3 - BASIC COMBINATIONAL CIRCUITS

SUMMARY

This project requires the design of simple combinational circuits. It provides some insight into intuitive minimization at the package level rather than the gate level.

ASSIGNMENT

I. Connect a NAND gate (1/4 of a 7400) to two switches as inputs and a light as output. Verify the logical behavior of the gate.

II. Repeat Assignment I with a NOR gate (1/4 of a 7402), an Exclusive OR (1/4 of a 7486), and an inverter (1/6 of a 7404).

III. For each of the following three circuit descriptions, design a circuit that uses only NAND gates (7400, 7410, 7420). Assume that speed (circuit delay) is of no concern and try to minimize the number of integrated circuit packages needed. Construct and test all three circuits. Assume that both complemented and uncomplemented inputs are available.

1) A 4-input, 1-output circuit with the following truth table:

w x y z	f	w x y z	f
0 0 0 0	0	1 0 0 0	1
0 0 0 1	0	1 0 0 1	1
0 0 1 0	0	1 0 1 0	1
0 0 1 1	0	1 0 1 1	1
0 1 0 0	1	1 1 0 0	1
0 1 0 1	1	1 1 0 1	1
0 1 1 0	1	1 1 1 0	1
0 1 1 1	0	1 1 1 1	0

2) A 4-input, 1-output circuit that outputs a 1 when three or four of the four inputs are 1 (majority circuit).

3) A 4-input, 3-output circuit that compares two 2-bit unsigned numbers, and outputs a 1 on one of three output lines according to whether the first number is less than, equal to, or greater than the other number (comparator circuit).

IV. Design a 5-output circuit for all three functions above, using any ICs in the standard kit including MSI, and sharing ICs to minimize the total number of packages. For this part assume that only uncomplemented inputs are available.

PARTS

This project must be completed using only parts from the standard kit.

HINTS

Assignments III (1), (2), (3), and IV can be completed using only 1, 2, 4, and 4 packages respectively. But the 4-package solution to Assignment IV is tricky, so don't waste a lot of time if it doesn't come to you in a half hour or so.

OPTIONS

Change the output encoding of the circuit for Assignment III (3) to allow a 3-package NAND implementation.

REFERENCES

The best exposition of techniques for logic design at the package level is given by Blakeslee [1975]. Discussions and examples of logic design with MSI can also be found in Gschwind and McCluskey [1975], Hill and Peterson [1974], Fairchild [1973], and Morris and Miller [1971]. Many texts describe classical combinational logic design, including McCluskey [1965], Kohavi [1970], Klir [1972], Kohonen [1972], Mano [1972], Peatman [1972], and Barna and Porat [1973].

A4 - BASIC SEQUENTIAL CIRCUITS

SUMMARY

The designs of simple counters and sequential machines are carried out in this project. The problem of contact bounce in switches is illustrated.

ASSIGNMENT

I.

1) Wire up the circuit shown in Figure A4-1. Use (debounced) pushbuttons for the CLK and CLR inputs and display the Q outputs of the flip-flops using four lamps. If it isn't already obvious, it should be apparent after you operate the clock button a few times that this is a 4-bit binary counter.

2) Now use an undebounced switch for the CLK input. The counter will now count very strangely because of contact bounce. Assuming that the contact bounces fewer than 16 times for each operation of the switch, take 10 samples and find the average number of contact bounces.

3) Show how an S-R flip-flop can be used to eliminate contact bounce.

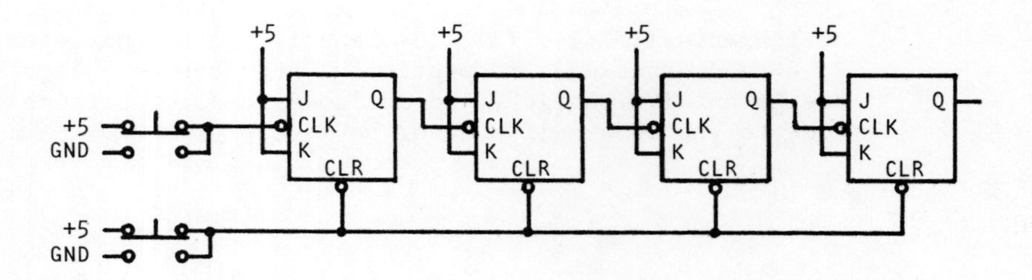

FIGURE A4-1 Ripple counter.

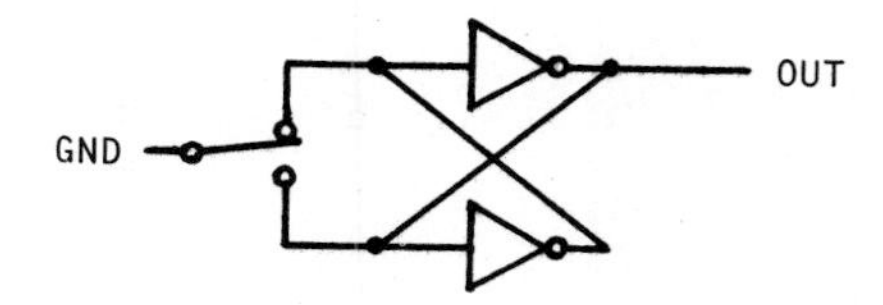

FIGURE A4-2 Switch debounce circuit.

4) The circuit of Figure A4-2 is sometimes used for debouncing, so that one TTL inverter (7404) package can debounce three switches, as opposed to one NAND (7400) package for two switches as in part (3). Explain the operation of this circuit. What assumptions have to be made for this circuit to work? Are they reasonable assumptions? (For new designs it is better to use circuits such as 74279 quadruple $\overline{S}$-$\overline{R}$ latches for debouncing four switches with one package).

II. Design a sequential circuit using only S-R flip-flops (Figure A5-1) and NAND gates. The circuit should have two inputs x and y, and a single output Z. Assume that only one of the input variables changes at a time. The output may change only in response to an input change. The output should be 1 if the xy input sequence (00, 01, 11) has been received and 0 otherwise. An example of the machine's operation is shown in Figure A4-3.

III. A T (toggle) flip-flop may have a clock input T, an enable input G, and an output Q. The output changes value (toggles) if a clock pulse (edge) occurs and the flip-flop is enabled (G=1). Design a T flip-flop (a) using only a single J-K master/slave flip-flop and b) using an edge-triggered D flip-flop and NAND gates. Do the two designs behave the same for all input sequences?

PARTS

This project requires only parts in the standard kit.

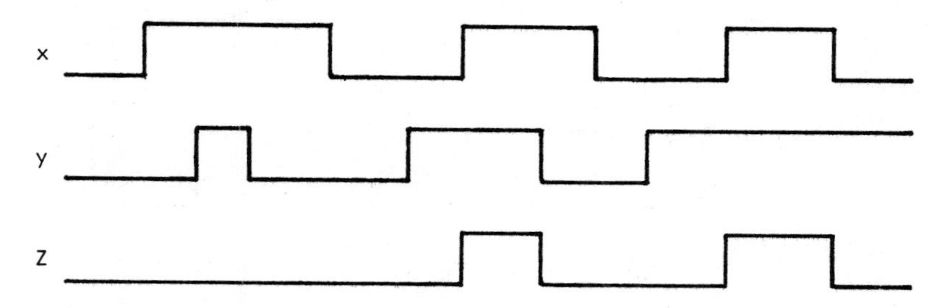

FIGURE A4-3 Example waveform.

REFERENCES

Discussions of sequential logic design can be found in each of the references for Project A3.

A5 - FLIP-FLOPS

SUMMARY

This project gives the NAND design of several important flip-flop types. It provides insight into the characteristics of each type, explaining concepts of latching, edge-triggering, and master/slave operation.

S-R FLIP-FLOP

An *S-R flip-flop* (or *S-R latch with asynchronous inputs*) consists of two cross-coupled NAND gates and possibly two inverters, as shown in Figure A5-1. S-R flip-flops can also be designed using cross-coupled NOR gates. A commercially available quadruple S-R flip-flop is the 74279.

CLOCKED S-R FLIP-FLOP

As shown in Figure A5-2, a *clocked S-R flip-flop* (or *S-R latch with synchronous inputs*) has an additional clock input so that the S and R inputs are active only when the clock is high. When the clock goes low, the state of flip-flop is "latched" and cannot change until the clock goes high again.

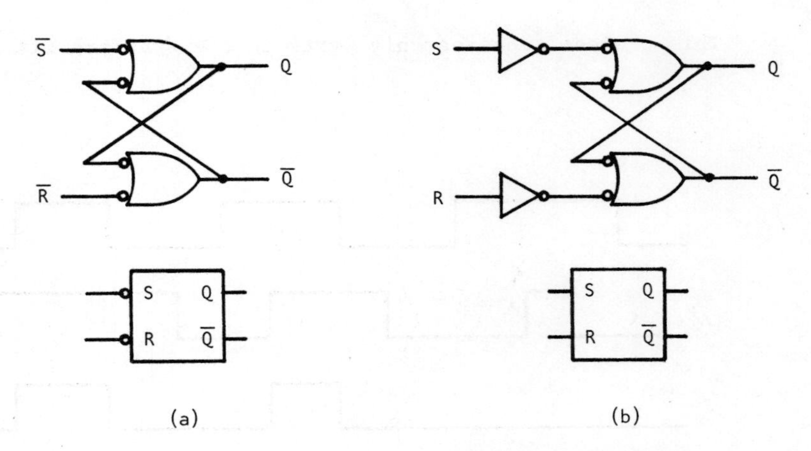

FIGURE A5-1 (a) $\overline{S}$-$\overline{R}$ flip-flop; (b) S-R flip-flop.

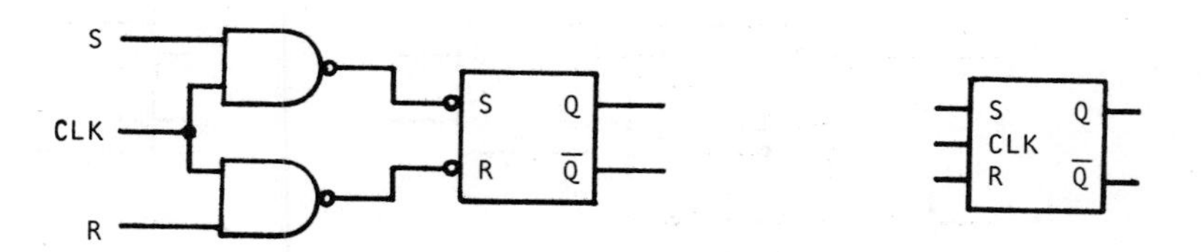

FIGURE A5-2 Clocked S-R flip-flop.

D LATCH

A *D latch* combines the S and R inputs of an S-R latch into one input, as indicated in Figure A5-3. When the clock is high, the output follows the D input, and when the clock goes low the state is latched. Commercially available D latches include 7475, 7477, 74100, and 74116.

D EDGE-TRIGGERED

An *edge-triggered D flip-flop* combines two D latches, as shown in Figure A5-4. The input latch is called the master and follows the input while the clock is low. When the clock goes high, the master is latched and its output is transferred to the second latch, called the slave. The slave output is seen by the user. Hence the edge-triggered D flip-flop senses the input data present at the rising edge of the clock and produces a corresponding output. The output can only change at the rising clock edge. Commercially available edge-triggered D flip-flops and registers include 7474, 74173, 74174, and 74175.

J-K MASTER/SLAVE

As shown in Figure A5-5, a *J-K master/slave flip-flop* is similar to an edge-triggered D flip-flop except that J and K inputs are provided for the master. Although the output can change only on the rising clock edge, a J-K master/slave flip-flop is not truly edge-triggered because the output does not always reflect the inputs present at the triggering edge (see Assignment III). Commercially available TTL master/slave J-K flip-flops include 7471, 7472, 7473, 7476, 7478, and 74107. All CMOS J-K flip-flops are true edge-triggered types.

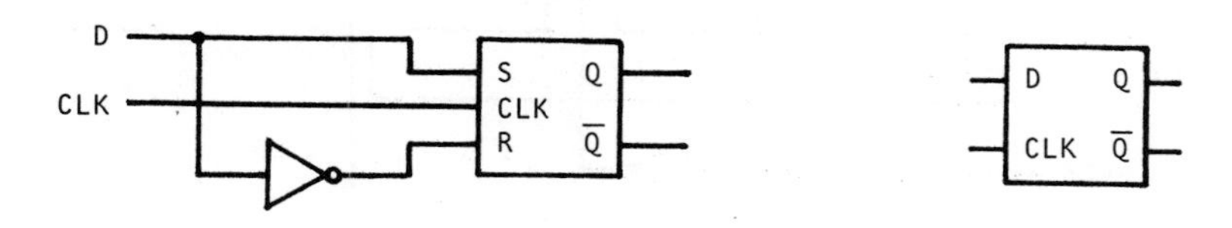

FIGURE A5-3 D latch.

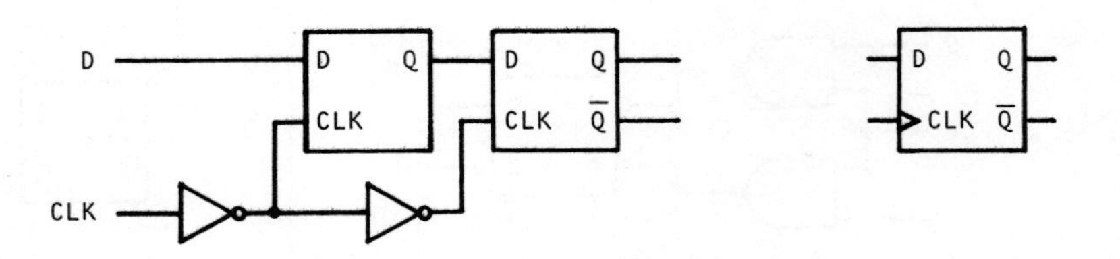

FIGURE A5-4 D edge-triggered.

J-K EDGE-TRIGGERED

Figure A5-6 shows one way of implementing a true *J-K edge-triggered flip-flop*, which produces an output that depends only on the input data present at the rising edge of the clock. Commercially available edge-triggered J-K flip-flops include 7470, 74101, 74102, 74103, 74106, 74108, 74109, 74112, 74113, and 74114.

ASYNCHRONOUS INPUTS

Asynchronous inputs can be added to any clocked flip-flop to allow the flip-flop to be forced to a state independent of the signals on the clock and synchronous inputs. An asynchronous input that forces Q to 1 is labeled *set* (S) or *preset* (PR), and an asynchronous input that forces Q to 0 is labeled *reset* (R) or *clear* (CLR).

CLOCK POLARITY

The clock polarity of any clocked flip-flop can be changed by inserting an inverter in the clock line. The presence of this inverter in the flip-flop is indicated by an inversion bubble on the clock input of the flip-flop symbol, as shown in Figure A5-7 for a *negative-edge-triggered* D-flip-flop. For latches the bubble indicates that the inputs are sensed while the clock is low, and for master/slave and edge-triggered devices it indicates that the output changes on the falling (negative) edge of the clock. A device without the inversion

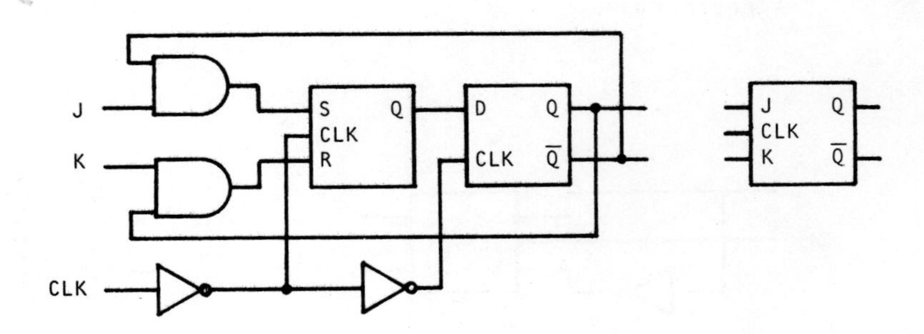

FIGURE A5-5 J-K master/slave.

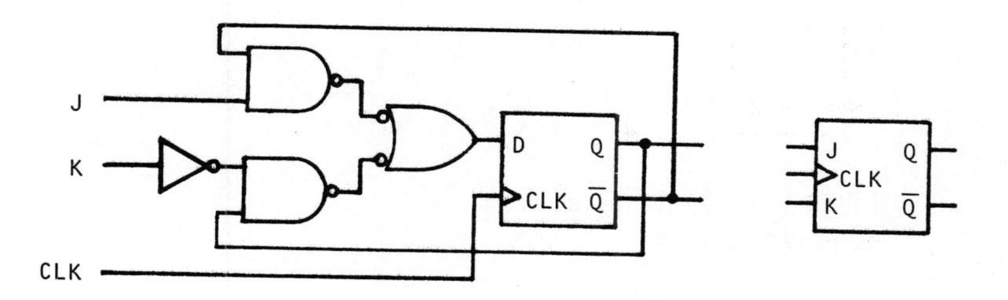

FIGURE A5-6 J-K edge-triggered.

bubble such as in Figure A5-4 is *positive-edge-triggered*. Note that devices with both positive and negative clocks have been included in the lists in the previous sections.

ASSIGNMENT

I. Design an S-R flip-flop using NOR gates.

II. Design and test each of the six flip-flop types above. Use NAND gates or 74279s if available to simplify the wiring. Try to minimize the number of gates for each flip-flop type.

III. Indicate on your schematics for each of the last five flip-flop types above what circuitry must be provided for asynchronous preset and clear inputs. These inputs would be used to set the flip-flop output independent of the clock.

IV. Compare the outputs of the J-K master/slave and edge-triggered flip-flops for the following input waveforms (both flip-flops should be initially cleared):

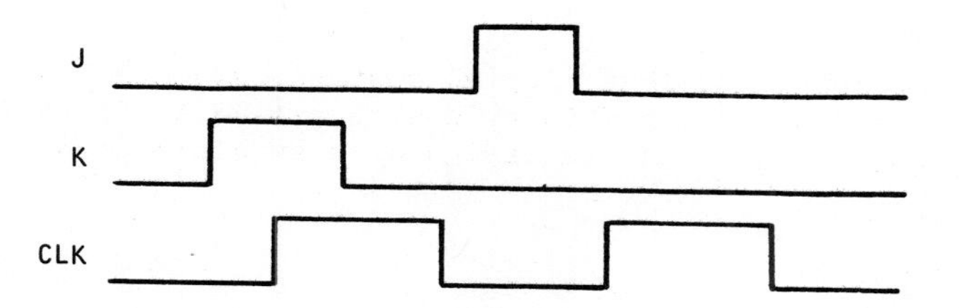

The behavior exhibited by the master/slave flip-flop is known as "ones catching." Devise and test an input waveform for which the master/slave flip-flop exhibits a similar behavior known as "zeroes catching."

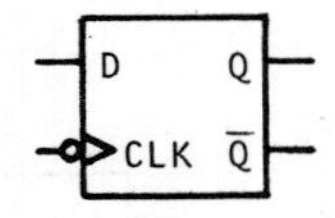

FIGURE A5-7 D negative-edge-triggered.

PARTS

This project can be completed using only parts in the standard kit.

HINTS

By drawing the NAND equivalents of the latches and S-R flip-flops it is easy to accomplish the minimization called for in Assignment II. The minimization consists of simply noticing which internal gates can be combined or eliminated when the flip-flops and latches are combined as shown.

REFERENCES

An excellent exposition of the differences between various types of flip-flops can be found in Gschwind and McCluskey [1975]. Flip-flop types are also described to some extent in Blakeslee [1975], Barna and Porat [1973], Kohonen [1972], Peatman [1972], Mano [1972], and Morris and Miller [1971].

A6 - COUNTERS

SUMMARY

This project describes ripple counters, synchronous serial and parallel counters, nonbinary counters, and shift register counters. The project requires the design of a specific counter system.

RIPPLE COUNTERS

Counters arranged so that the output of one flip-flop generates the clock input of the next higher stage are generally called *ripple counters*. A 4-bit ripple counter was shown in Figure A4-1. When a transition from, say, 0111 to 1000 occurs, the one-to-zero transition of the low-order three bits ripples from bit to bit. Since each flip-flop has a nonzero propagation delay, ripple counters have several distinct limitations that complicate the decoding of their states and place an upper limit on the number of stages that may be correctly decoded. Rather than dwell upon these limitations, we proceed directly to synchronous counters.

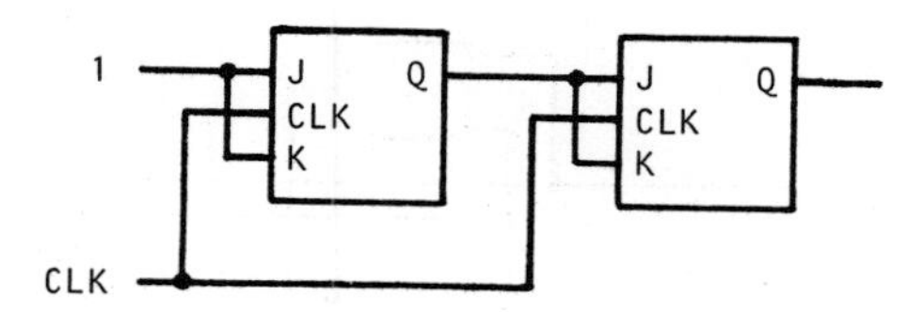

FIGURE A6-1 2-bit synchronous counter.

SYNCHRONOUS COUNTERS

In a *synchronous counter* each flip-flop is clocked by the same clock signal. Whether a flip-flop changes state in response to the clock is controlled by the synchronous enabling inputs, for example, the J and K inputs of a J-K flip-flop. The important feature of a synchronous counter is that the transitions of the individual flip-flops are synchronized to the master clock signal. In the synchronous transition from 0111 to 1000, for example, there is no rippling process; the bits switch simultaneously. The two-bit counter in Figure A6-1 is the simplest example of a synchronous counter. Notice that like its ripple counterpart no components other than flip-flops are required. In general synchronous counters require external gating and so they are more expensive than ripple counters. Nonetheless they are far more widely used.

There are two basic schemes for generating the enable J and K inputs. One of them is illustrated in the four-bit binary counter in Figure A6-2. Notice that the information to complement travels in a serial fashion from one stage to the next. This counting scheme is accordingly termed a *synchronous serial counter*.

If the J-K input information is formed in a parallel fashion, one has a *synchronous parallel counter* such as shown in Figure A6-3. Notice that while the number of required gate inputs per stage is constant in the serial case (two inputs per stage), in the parallel scheme the number of inputs increases linearly with the number of stages. For this added expense one gets the fastest possible synchronous counting circuit.

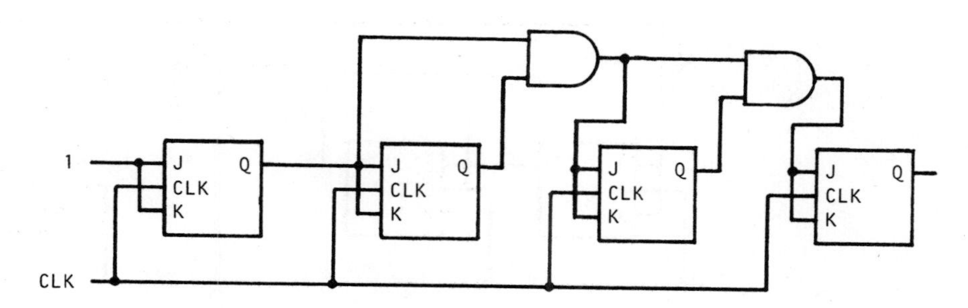

FIGURE A6-2 4-bit serial synchronous counter.

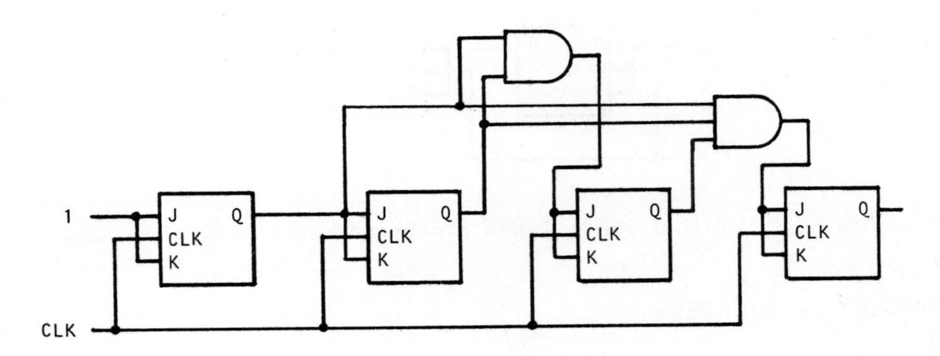

FIGURE A6-3 4-bit parallel synchronous counter.

NONBINARY MODULUS COUNTERS

It is often desirable to count with a modulus different from a power of two. The most common case of this need is a *BCD counter*, which counts in a binary fashion from zero to nine and then returns synchronously to zero. Given a synchronous counter, it is generally desirable that all its transitions are synchronous, including the return-to-zero transition, rather than simply forcing a zero via the asynchronous clear inputs. A three-bit binary modulo 5 counter is shown in Figure A6-4.

NONNUMERICAL ORDER

A binary sequence in ascending numerical order is only one way in which a counter may count. A Gray code is another. One may specify any arbitrary sequence and design a counting circuit to step through it. In many situations the counting sequence is not specified, and only the number of states is defined. In these situations one is free to choose a counting sequence that optimizes some performance measure such as cost, speed, or physical size.

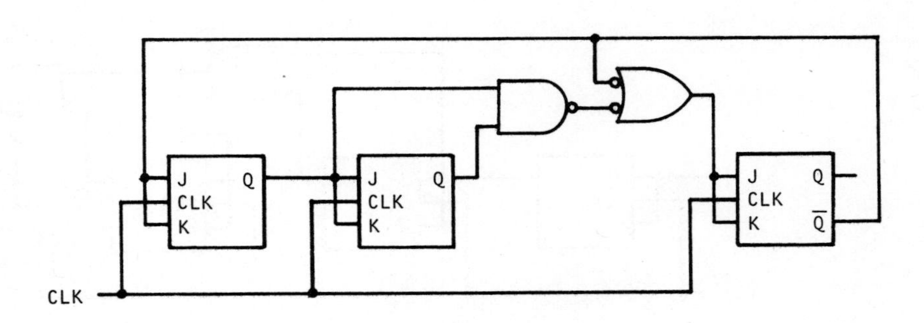

FIGURE A6-4 Modulo 5 counter.

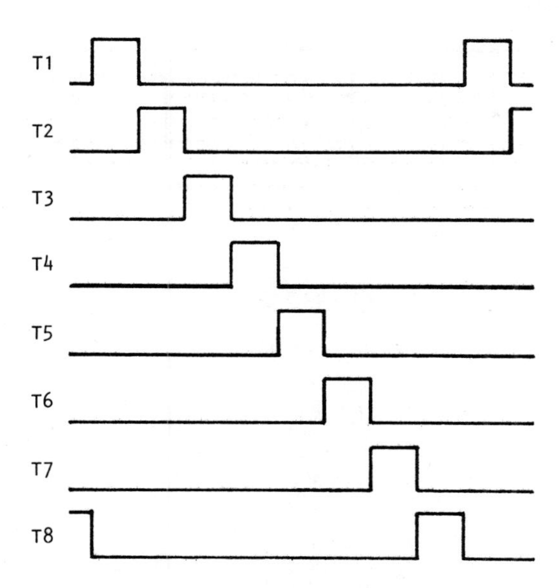

FIGURE A6-5 An eight-phase clock.

SHIFT REGISTER COUNTERS

Suppose one is required to generate the set of timing pulses shown in Figure A6-5. Although it is possible to decode them from the states of a 3-bit binary counter, one can instead use an 8-bit circular shift register loaded with a single circulating 1. This circuit, called a *ring counter*, requires more flip-flops but does not need any external gating to produce the desired timing signals.

A compromise between an n-bit binary counter and a 2^n-stage ring counter is a circuit variously called a *switchtail ring counter*, *switchback counter*, *Moebius counter*, or *Johnson counter*. An n-stage Johnson counter has $2n$ states. A 4-stage counter and the gating needed to produce the timing pulses of Figure A6-5 are shown in Figure A6-6.

An n-stage Johnson counter requires only a 2-input gate to decode each of its states while an n-bit binary counter requires n-input gates. Furthermore, as the Johnson counter shifts from one state to the next, there is never a decoding error. In a binary counter false decoding may occur, for example, during the 0011 to 0100 transition if the counter briefly passes through the state 0010. The decoded outputs of the Johnson counter are clean and free from "glitches."

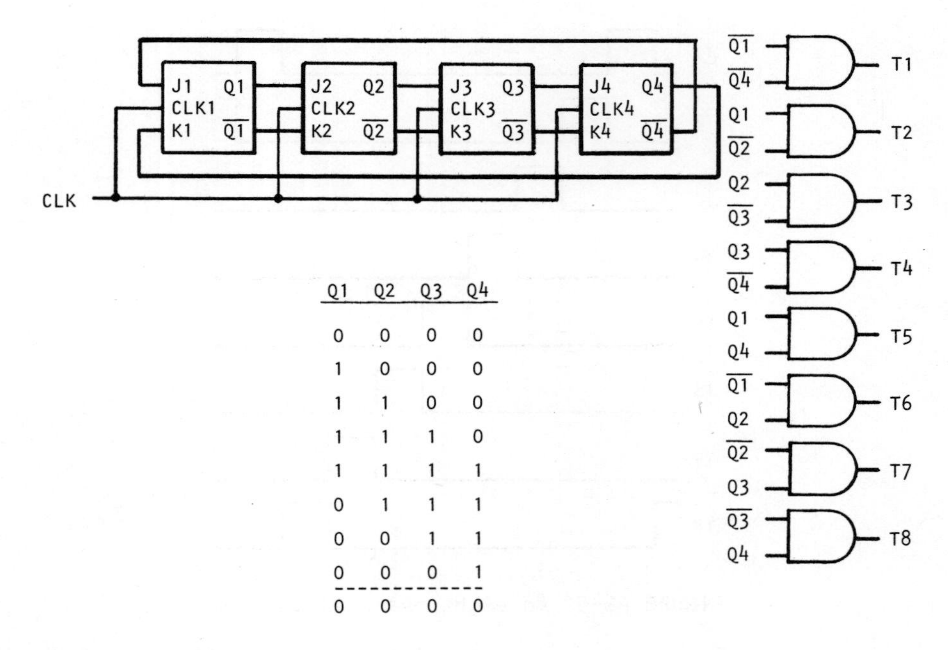

Q1	Q2	Q3	Q4
0	0	0	0
1	0	0	0
1	1	0	0
1	1	1	0
1	1	1	1
0	1	1	1
0	0	1	1
0	0	0	1
0	0	0	0

FIGURE A6-6 4-stage Johnson counter.

INITIALIZATION

For counting circuits that do not use all the possible states, there will be illegal states, and one must in general provide some mechanism to detect or correct these unallowed states. Most shift register counters have illegal states that can occur when the circuit is powered up or experiences a transient failure. Hence the circuit must have some way to keep an illegal state from recirculating forever. For the Johnson counter shown in Figure A6-6, such a correction scheme is shown in Figure A6-7.

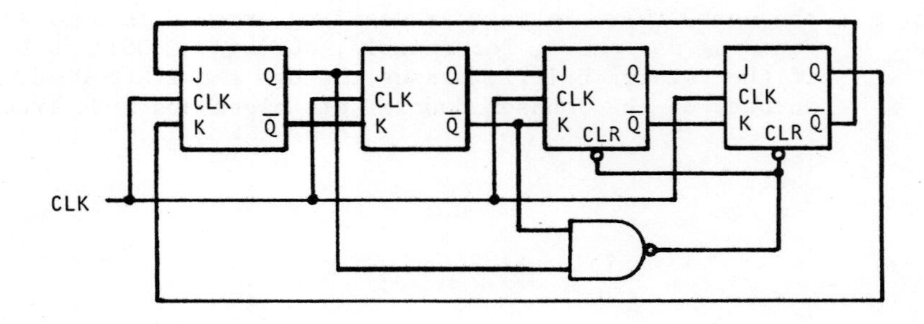

FIGURE A6-7 Self-correcting Johnson counter.

ASSIGNMENT

I. Show how to add an enable input E to the synchronous counters of Figures A6-2 and A6-3. When E is 1, the counters should count at each clock pulse, and when E is 0 no count should take place. The incoming clock should still go directly to each flip-flop. What constraints are placed on changing E if master/slave J-K flip-flops are used?

II. You are to design a circuit with the following characteristics:

1) Divide a 60 Hz clock from a pulse generator down to 1 Hz. You should use a synchronous binary counter that has exactly 60 states.

2) Use the 1 Hz output from the divide-by-60 counter to drive a 5-stage Johnson counter with 10 allowable states. You should insure that any illegal sequence will be corrected.

3) Decode these 10 states to produce a 1-out-of-10 code that you apply to 10 lamps so just the first lamp is lit, then just its neighbor, and so on, wrapping around from one end back to the first. (If it is more convenient, you may use a 9-out-of-10 code in which only one lamp is off at all times.)

PARTS

Use only NAND gates and flip-flops from the standard kit. Do not use any MSI counters.

REFERENCES

Discussion of various types of counters appear in Gschwind and McCluskey [1975], Blakeslee [1975], Barna and Porat [1973], Kohonen [1972], Peatman [1972], Mano [1972], and Morris and Miller [1971].

A7 - SHIFT REGISTERS

SUMMARY

This project introduces shift registers and some of their applications. The designs of a self-starting ring counter, a maximum-length sequence generator, and a variable-length shift register are required.

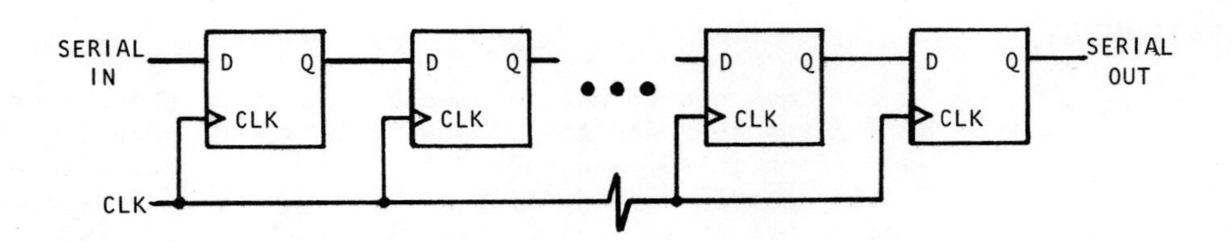

FIGURE A7-1 Shift register.

SHIFT REGISTER BASICS

A shift register is an *n*-bit register with provision for shifting its stored data by one position at each clock pulse. As shown in Figure A7-1, a new bit to be shifted into one end must be specified, and the bit shifted off the other end is lost unless it is saved externally. Although Figure A7-1 shows a right-shift register, the same register can obviously be used for left shifts simply by reversing the sense of the bits. Most shift registers have provision for shifting only in one direction, but some have a control input that allows either left or right shifting to be specified at each clock.

One way to load an *n*-bit shift register is to shift the bits into the register one at a time using the serial input. Some shift registers also have parallel inputs that can be used to load all *n* bits in one clock cycle. The output of a shift register can be observed one bit at a time at the serial output, but some also have parallel outputs for observing all *n* bits at once. Standard 7400-series shift registers may be classified by their type of inputs and outputs: *serial-in, serial-out* (7491); *serial-in, parallel-out* (74164); *parallel-in, serial-out* (7494, 74165, 74166); and *parallel-in, parallel-out* (7495, 7496, 7499, 74178, 74179, 74194, 74195, 74198, 74199, 74295). Most of the registers shift in one direction only, except the 7499, 74194, 74198, and 74295, which are bidirectional.

DATA CONVERSION

A primary application of shift registers is to convert data from parallel to serial format or vice versa. To convert *n* bits from parallel to serial they are loaded in parallel into an *n*-bit parallel-in, serial-out shift register. Then they are serially shifted out.. Serial-to-parallel conversion uses a serial-in, parallel-out shift register in the obvious manner.

COUNTERS

Two types of shift register counters - ring counters and Johnson counters - were discussed in Project A6. The main advantages of shift register counters are simplicity of decoding and freedom from decoding transients.

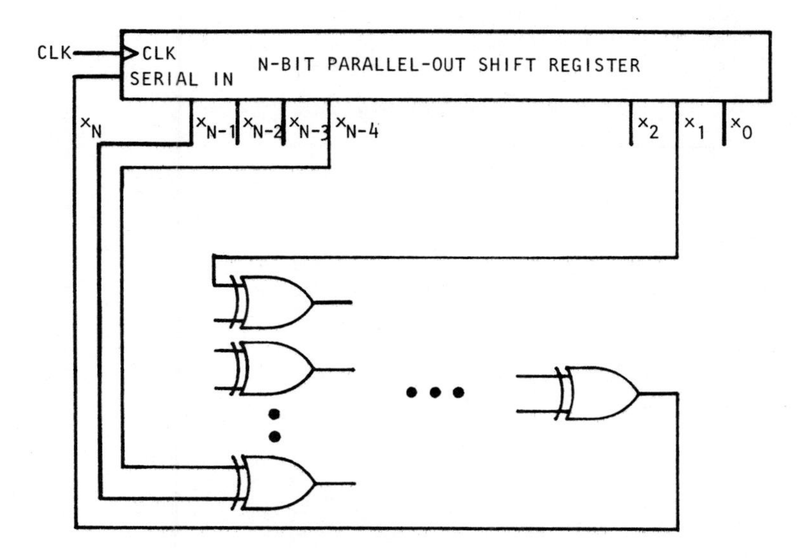

FIGURE A7-2 Maximum-length sequence generator.

MAXIMUM LENGTH SEQUENCES

An n-bit binary counter has 2^n different states. An n-bit shift register circuit with almost as many states can be designed with less circuitry than an n-bit synchronous counter, and it will operate just as fast. When started in a nonzero state it will visit all 2^n-1 nonzero states before returning to the starting state. Such a circuit is called a *maximum length sequence generator*, and it has the general form shown in Figure A7-2. The input of the serial-in, parallel-out shift register is connected to the modulo-2 sum (exclusive OR) of selected shift register output bits. The output bits used correspond to the nonzero coefficients of a primitive polynomial of order n (see References). For any register length n there are several choices of output bits to be summed which result in a maximum length sequence; Table A7-1 gives a choice of the minimum number of output bits needed for selected values of n..

Although a sequence of states generated by the above scheme is not in numerical order, such sequences are useful in many applications. The serial output of the sequence generator produces a pseudo-random bit stream with a period of 2^n-1. Such a bit stream is useful in applications such as cryptographic encoding and improving signal-to-noise ratios in digital communications. The parallel output of the shift register can be used to obtain all 2^n-1 nonzero n-bit numbers in pseudo-random order; hence the generator acts as a pseudo-random number generator. However, there is a high serial correlation between successive outputs since they are obtained by simple shifts.

TABLE A7-1 Maximum-length Sequence Equations

n	Feedback Equation
2	$x_2 = x_1 \oplus x_0$
3	$x_3 = x_1 \oplus x_0$
4	$x_4 = x_1 \oplus x_0$
5	$x_5 = x_2 \oplus x_0$
6	$x_6 = x_1 \oplus x_0$
7	$x_7 = x_3 \oplus x_0$
8	$x_8 = x_4 \oplus x_3 \oplus x_2 \oplus x_0$
9	$x_9 = x_4 \oplus x_0$
10	$x_{10} = x_3 \oplus x_0$
11	$x_{11} = x_2 \oplus x_0$
12	$x_{12} = x_6 \oplus x_4 \oplus x_1 \oplus x_0$
16	$x_{16} = x_5 \oplus x_4 \oplus x_3 \oplus x_0$
20	$x_{20} = x_3 \oplus x_0$
24	$x_{24} = x_7 \oplus x_2 \oplus x_1 \oplus x_0$
28	$x_{28} = x_3 \oplus x_0$
32	$x_{32} = x_{22} \oplus x_2 \oplus x_1 \oplus x_0$

SERIAL MEMORY

Large serial-in, serial-out shift registers are often used in applications that require data to be cyclically accessed in some fixed order. The most common of such applications are raster-scan CRT terminals, in which shift registers are used to buffer a line of text to be displayed. Very large CCD (charge-coupled device) and magnetic-bubble shift registers are also being offered as IC replacements for mechanical serial memories such as magnetic discs and drums.

ASSIGNMENT

I.

1) Design an 8-bit ring counter (with a single circulating 1) using two 4-bit MSI shift registers. Display the output using eight lamps. Provide a reset push button to initialize the circuit to the 10000000 state. Do not use the parallel inputs of MSI shift registers if they have a clear input.

2) Design self-starting and self-correcting circuitry for the above counter. With this circuitry the counter should automatically return to a valid state in a few cycles regardless of its starting state, and without the aid of a reset button.

3) Design a self-starting and self-correcting "inverse" ring counter, that is, a counter with a single circulating 0.

4) Build a 3-bit ripple counter using 3 J-K flip-flops and use a 3-input NAND gate to decode the 000 state. Connect a pulse generator to the ripple counter and the counter of part (3). Using a high-speed oscilloscope ($\geq$ 15 MHz), compare the decoded 000 output and the low-order shift register bit.

II.

1) Design an 8-bit maximum length sequence generator. The register should cycle through 255 different nonzero states. If you have available a logic state analyzer or a pulse burst generator, verify that the initial state is repeated after 255 clock pulses, and that it is not repeated after 3, 5, or 17 clock pulses. Whether you have the test equipment or not, explain why this procedure is sufficient to verify that the period of generator is exactly 255.

2) If started in the 00...0 state, the sequence generator above will remain in the 00...0 state forever. Yet most shift registers have provision initializing the register to the 00...0 state. Design an 8-bit maximum length sequence generator that includes the 00...0 state, and when initialized to 00...0 counts through 255 states excluding the 11...1 state.

III. Design a 4-bit variable-length ring counter. The circuit should have two toggle switches to select a length of 1, 2, 3, or 4 bits for the register. For any length the unused bits should always be 0. Thus the state sequences required are as follows

length = 1	length = 2	length = 3	length = 4
1000	1000	1000	1000
1000	0100	0100	0100
1000	1000	0010	0010
1000	0100	1000	0001
1000	1000	0100	1000
...	...	...	...

Your circuit should be self-starting and self-correcting for any length sequence.

PARTS

This project can be completed using only parts in the standard kit.

HINTS

Assignment I(3) can be accomplished simply by changing the self-starting circuitry of I(2). One way to solve Assignment II(2) is to conceptually invert all of the input and output bits of the shift register of II(1).

REFERENCES

Discussions of shift registers and shift register counters can be found in most of the references for Project A6. Maximum-length sequence generators and random number generators are described in Stone [1973] and Peterson and Weldon [1972].

A8 - BINARY RATE MULTIPLIERS

SUMMARY

The principles and uses of binary rate multipliers (BRMs) are described in the project. A BRM is used to construct a simple digital integrator. The integrator is connected in a loop to generate exponential functions.

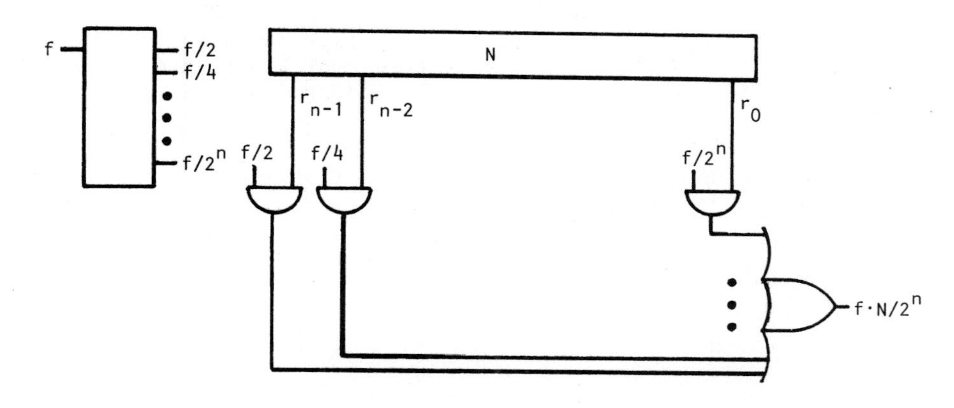

FIGURE A8-1 Binary rate multiplier.

BRM PRINCIPLES

A *binary rate multiplier (BRM)* is a sequential circuit with a clock input CLK and n rate inputs $r_0, r_1, \ldots, r_{n-1}$. The rate inputs are the binary representation of a number $N = r_0 2^0 + r_1 2^1 + \ldots + r_{n-1} 2^{n-1}$. The binary rate multiplier has a single output Z. If the input rate is N, then N output pulses appear at Z for every 2^n input pulses at at CLK. Hence, if f is the input frequency, the output frequency is $f \cdot N/2^n$; the input frequency is multiplied by $N/2^n$.

The BRM accomplishes frequency multiplication as shown in Figure A8-1. Each bit r_i of N is used as the input to an AND gate whose other input is the input frequency f divided by a power of 2, namely 2^{n-i}. Hence, if r_i is 1, the corresponding AND gate output will have $2^n/2^{n-i} = 2^i$ output pulses for ever 2^n input pulses. If r_i is 0, the AND gate has no output pulses. The AND gate outputs are collected in an OR gate to produce the output Z. As long as none of the AND gate input pulses occur simultaneously, the OR gate transmits the sum of the input pulses it receives, or $r_0 2^0 + r_1 2^1 + \ldots + r_{n-1} 2^{n-1}$ output pulses for every 2^n input pulses.

The frequencies $f/2, f/4, \ldots, f/2^n$ can be generated using a simple n-bit binary counter whose input frequency is f. Noncoincidence of pulses is obtained by emitting a pulse at a particular stage of the counter only when that state changes from 0 to 1. Table A8-1 shows the behavior of a 3-bit counter for generating $f/2$, $f/4$, and $f/8$. It is clear that exactly one stage has a 0-to-1 change for each count, except for the 111 to 000 count, which has none. It is also clear that the correct frequency is produced at each stage.

TABLE A8-1 BRM Frequency Generation

State	Q_0 Q_1 Q_2	$f/2$	$f/4$	$f/8$
0	0 0 0	—		
1	1 0 0		—	
2	0 1 0	—		
3	1 1 0			—
4	0 0 1	—		
5	1 0 1		—	
6	0 1 1	—		
7	1 1 1			
		r_2	r_1	r_0

COMPARISON WITH REGISTER/ADDER SYSTEMS

The operation of an n-bit BRM can be compared with that of an n-bit register and adder system shown in Figure A8-2. The adder output is loaded into the register at each clock pulse. The inputs of the are an external rate input and the register output. Hence at each clock pulse the register is increased by the amount of the rate input. From time to time, the adder will overflow and produce an output carry at C. A little reflection shows that if the rate input is N, then N output carries will be produced for every 2^n clock pulses, just as in a BRM.

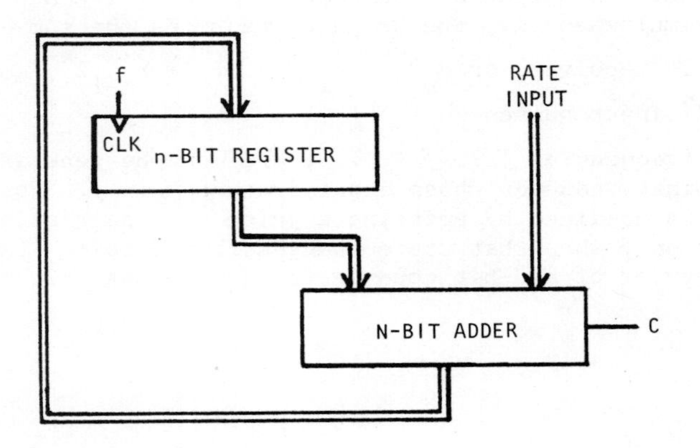

FIGURE A8-2 Register/adder system.

Hence we can use the register/adder system of Figure A8-2 to simulate a BRM. Conversely, we can use a BRM to simulate the register/adder system. The register/adder system and the BRM are equivalent in systems where the rate inputs are constant. However, if the rate inputs change during operation then the register/adder and the BRM are not equivalent. The state of the register/adder system is the accumulated sum of the rate inputs present at each clock pulse. The state of the BRM system, on the other hand, depends only on the number of clock pulses received and is independent of the rate inputs. The interpretation of the BRM state changes whenever the rate inputs change. Consider a system in which the rate inputs are alternating between 10...0 and 01...1. In the register/adder system, an output carry will be produced on approximately every second input pulse. The behavior of the BRM output is radically different, as shown in Assignment I(3). This is an extreme example. In cases where the rate inputs do not change too often, the BRM gives a reasonably accurate simulation of a register/adder system.

AN INTEGRATED CIRCUIT BRM

In this project we will be using the Texas Instruments 7497 synchronous 6-bit binary rate multiplier. Referring to the manufacturer's schematic (Figure A8-3) it can be seen that the structure of the 7497 is similar to that described earlier. A 6-bit synchronous counter is used to produce the submultiples of f. Noncoincidence of pulses is obtained by using AND gates to recognize the states for which output pulses are to be produced. The same AND gates are used for the rate inputs and for the clock to achieve synchronism. Notice that the Z

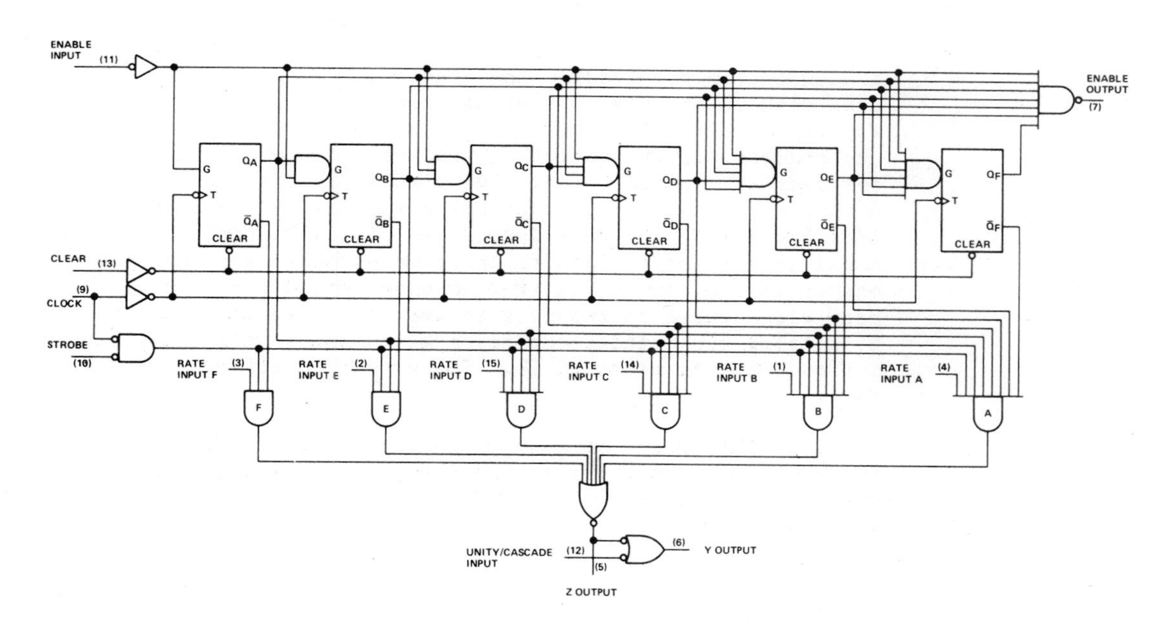

FIGURE A8-3 7497 6-bit BRM (courtesy of Texas Instruments).

output of the 7497 is normally high. The counter changes state on the rising edge of the clock pulse. The Z output goes low when the clock goes low if an output pulse is to be produced, otherwise it remains high. The 7497 has a clear input and some extra inputs and outputs that are useful in cascading 7497s to make larger BRMs.

ASSIGNMENT

I. Wire up a 7497 using push buttons for the clock and clear, toggle switches for the six rate inputs, and a lamp for the Z output.

1) For each of the numbers 1, 5, 9, and 63 do the following. Enter the binary representation of the number in the BRM rate inputs, clear the BRM, apply 64 input pulses, and verify that the correct number of output pulses is produced.

2) Clear the BRM, and set 44 as the rate input. Apply 16 input pulses. Sketch the resulting waveforms of the clock and the Z output. On the same figure, sketch the waveforms that are present at the outputs of the F, E, D, and C AND gates internal to the 7497.

3) Clear the BRM, and set 011111 as the rate input. Apply one clock pulse and set 100000 as the rate input. Apply another clock pulse and apply 011111 again as the rate input. Continue alternating the rate input between 011111 and 100000, and observe the number of output pulses produced. Repeat the experiment, clearing the BRM but starting first with 100000 as the rate input. Explain the behavior observed in both cases.

II. Wire up an 8-bit up/down counter. Use 8 toggle switches and a push button for loading the counter, and check out its operation. Then connect the six most significant bits of the counter as the rate inputs of a 7497 BRM. Connect a push button to the BRM clock input and connect the BRM output so that the counter is incremented once for each BRM output pulse. The circuit that you have constructed is a simple digital integrator that simulates the exponential function. (See Project D5 for more details on digital integrators.) The initial condition is set by loading the counter and clearing the BRM. The output of the counter is then an exponential function of time, where each clock pulse corresponds to one time quantum. Test the circuit by setting the initial condition at 64 and plotting the output for $t=0$ to $t=300$ in steps of 10. The output function should be exponential.

III. Change the circuit so that the counter is decremented once for each BRM output pulse. Load 128 as the initial condition and plot the output for $t=0$ to $t=300$ in steps of 10. What is the output function? Explain.

PARTS

You will need a 7497 BRM and two 4-bit up/down counters (74191 or 74193) to complete this project.

HINTS

Use a negative clock for the BRM in Assignment I. That is, the clock input of the 7497 should go low when the clock push button is depressed.

OPTIONS

Display the output functions of Assignments II and III using a DAC and an oscilloscope (see Project A9). To produce the display you will need to use a pulse generator for the BRM clock input and devise a way to reset the digital integrator with the initial condition when it overflows or underflows. Then build an analog circuit that generates exponentials and compare the shape of its output with your integrator's.

REFERENCES

At the time of this writing, the only 7400-series BRMs manufactured were by Texas Instruments [Texas Instruments, 1972]. A CMOS BRM (4089) is also available [RCA, 1975]. The use of BRMs to perform digital integration is discussed briefly in Sizer [1968].

A9 - DIGITAL-TO-ANALOG CONVERTERS AND OSCILLOSCOPE DISPLAYS

SUMMARY

This project describes the operation of digital-to-analog converters (DACs). The output of an 8-bit counter is connected to a DAC to produce an analog sawtooth function. The use of DACs to produce X-Y oscilloscope displays is also investigated.

DAC PRINCIPLES

An n-bit *digital-to-analog converter (DAC)* has n binary input lines, $x_0, \ldots, x_{n-1}$, and an analog output line V_{out}. The analog output is proportional to the weighted sum of the binary inputs, that is, $V_{out} = k(x_0 \cdot 2^0 + x_1 \cdot 2^1 + \ldots + x_{n-1} \cdot 2^{n-1})$. In some DACs, called "multiplying DACs," k is not constant but rather it is proportional to an external reference voltage V_{ref}. Hence a multiplying DAC gives an output proportional to the product of the binary input number and V_{ref}.

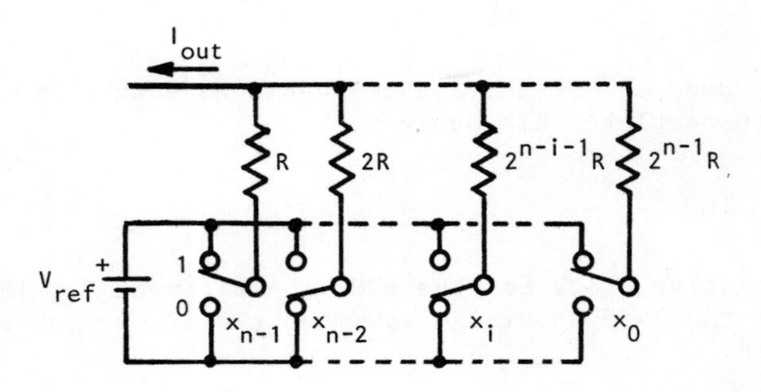

FIGURE A9-1 DAC weighted resistor network.

The principle of DAC operation is simple. Each of the n binary inputs is used as a switch to control a current source proportional to the weight of that input. If the input is 1 the current flows, and if the input is 0 no current flows. The currents from the n switches are summed and amplified to produce a voltage or current output proportional to the binary input number.

Figure A9-1 is a simplified circuit diagram of an n-bit DAC. The eight values are chosen so that the current produced when $x_i = 1$ is $V_{ref}/(R\cdot 2^{n-i-1}) = [V_{ref}/(R\cdot 2^{n-1})]\cdot 2^i$, and the current is proportional to the weight of x_i as required.

One disadvantage of the weighted resistor network used in Figure A9-1 is that a large range of resistances must be provided, with the largest resistor having 2^{n-1} times the resistance of the smallest. Also, different tolerances are required for each resistor for accurate operation. For example a 10% variation in the resistor for the least significant bit causes an error of only $(10/2^n)$% of the maximum DAC output; but a variation of 10% in the most significant resistor causes an error of (10/2)%! To obtain accuracy of 1 part in 1000 for

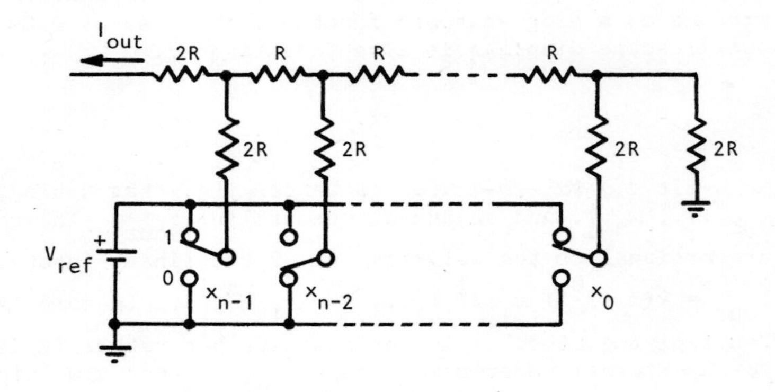

FIGURE A9-2 DAC ladder network.

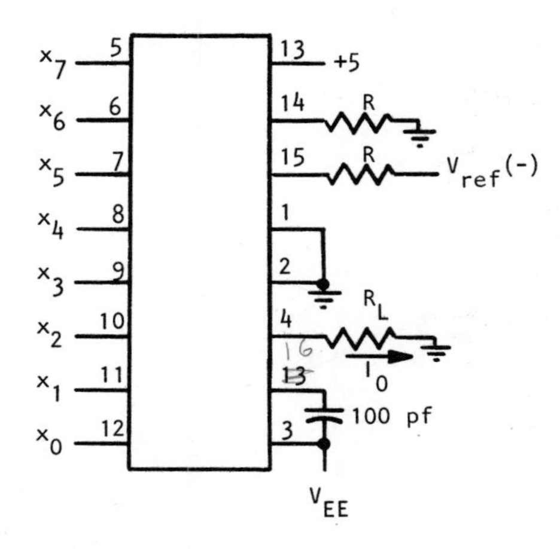

FIGURE A9-3 MC1408L-8 with negative reference.

a 10-bit DAC, the resistor for the least significant bit may have a tolerance of ±50%, while the most significant resistor must have a tolerance of ±.05%! Such absolute tolerances are impossible to achieve in integrated circuits.

The above difficulties are usually overcome by using a ladder network as shown in Figure A9-2. Using simple circuit analysis (Ohm's law and the principle of superposition) one can show that each branch of the network contributes an output current proportional to its weight. Only two different resistor values are needed. While it is still impossible to precisely set the absolute resistor values in IC fabrication, their relative values can be made quite accurate over the small 2-to-1 range. With the ladder network, DAC accuracy comparable to relative resistor accuracy is obtained.

It is possible to design DACs that accept a number of 4-bit BCD digits rather than binary inputs. In such DACs the resistor values are chosen to produce currents proportional to the BCD weights, for example, 80, 40, 20, 10, 8, 4, 2, 1 for a 2-digit DAC. Both weighted and ladder networks for BCD conversion can be designed.

AN INTEGRATED CIRCUIT EIGHT-BIT DAC

An inexpensive 8-bit multiplying DAC is the Motorola MC1408L-8. This circuit has TTL compatible inputs and is supplied in a 16-pin package. It can use either a positive or negative reference voltage and produces an output of the opposite polarity.

A connection diagram for the MC1408L-8 for negative reference voltages is shown in Figure A9-3. The output current is given by $I_o = -[V_{ref}/R]\cdot[\sum_{0\leq i\leq 7} x_i 2^i]/256$. The maximum allowable value of $[V_{ref}/R]$ and hence I_o is 2 ma. Also, R_L must be chosen so that V_{out}

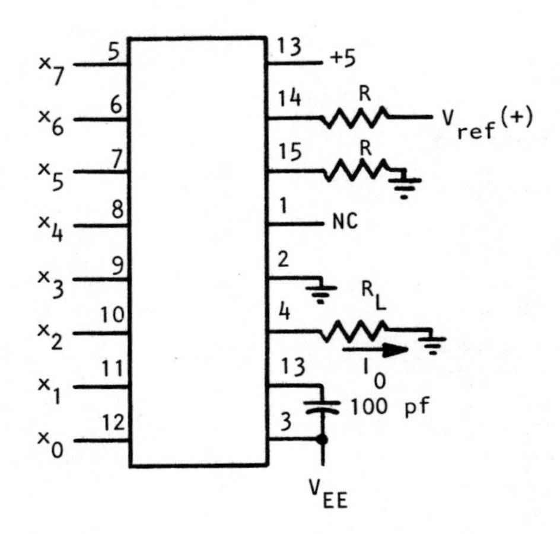

FIGURE A9-4 MC1408L-8 with positive reference.

does not exceed .5 volt. Typical values are V_{EE} = -5 volt, V_{ref} = -2 volt, R = 1K, R_L = 250, and $R_{out(max)}$ = .5 volt.

For positive reference voltages, the MC1408L-8 may be connected as shown in Figure A9-4. The output current is given as before and is limited to 2 ma. However, V_{out} may be as low as -5 volts; the only restriction is that V_{EE} must be at least 4.5 volts more negative than V_{out}. Typical values are V_{EE} = -12 volts, V_{ref} = 2 volts, R = 1K, R_L = 2.5K, and $V_{out(min)}$ = -5 volts.

Bipolar reference voltages may be handled by a connection similar to Figure A9-3, except that the pin 14 resistor should be connected to a positive reference voltage equal to the peak positive reference input at pin 15.

The reference voltage can be obtained from the logic supply or the V_{EE} supply, but care must be taken to avoid coupling power supply noise into the reference input. The reference resistor should be decoupled by connecting it to the supply through another resistor and bypassing the junction of the two resistors with 0.1 uf to ground.

OSCILLOSCOPE DISPLAYS

Most oscilloscopes have an X-Y mode in which signals may be connected directly to the vertical and horizontal amplifiers of the scope to determine the X-Y position of the beam. With this mode the user can move the beam to produce any desired display. To avoid flicker, the displayed image must be repeated ("refreshed") about 60 times per second. The scopes may also have a Z input that can be used to blank the display when unwanted X-Y inputs are present.

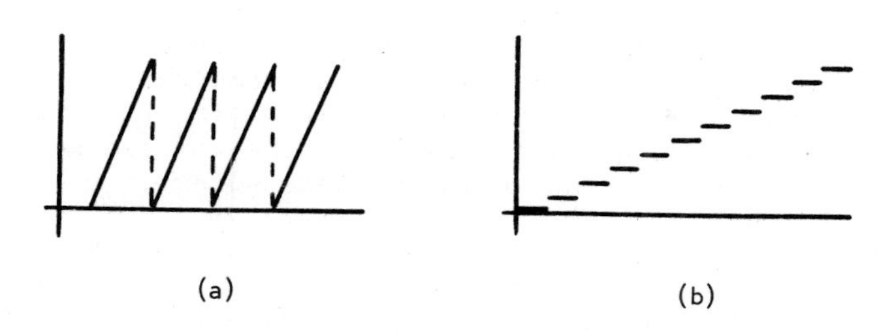

FIGURE A9-5 DAC waveforms.

ASSIGNMENT

I. Wire up an 8-bit binary counter. Connect a pulse generator to the clock input of the counter and use an oscilloscope to verify that the output of the low order counter bits is half the clock frequency, and that each succeeding bit is half the frequency of the previous bit.

II. Hook up an 8-bit DAC; you may follow Figure A9-3 or A9-4 for an MC1408L-8. Check all of the DAC wiring and the values of the V_{EE} and V_{ref} supplies before connecting them to the DAC - analog circuits such as DACs are not always as tolerant of abuse as TTL. With the TTL data inputs floating, the DAC output voltage corresponding to an input of 11111111 should be obtained. The output voltage of the MC1408L-8 can be varied by changing V_{ref} or the reference resistor (R), but the reference current (V_{ref}/R) must not be allowed to exceed 2 ma.

III. Connect the eight counter output bits to the DAC. With the counter running at a reasonable speed (say 100 KHz), observe the DAC output using an oscilloscope. The output waveform should look like Figure A9-5a for a positive DAC output, or inverted for a negative DAC output. Answer the following questions:

1) What is the DAC output voltage when the digital input word is 00000000? 11111111? 01000000?

2) With the counter running at 100 KHz, how long does it take for one complete cycle of the DAC output waveform? At what speed would the clock have to run for the DAC output waveform to make 60 cycles per second?

Increase the sweep rate and the sensitivity of the scope vertical input to "stretch out" the DAC waveform. You should be able to observe the transitions between DAC output states, as shown in Figure A9-5b.

3) Sketch the waveform that follows the 11111111 to 00000000 transition, and the waveform for the next few clocks. Compare this waveform with the behavior of typical transitions and explain.

4) Speed up the clock to 500 KHz. Stretch out the waveform some more. What is the settling time of typical DAC transitions? The settling time is the time required for the DAC voltage to reach and stay within 1/2 LSB of its final value, where LSB ("least significant bit") is the voltage between DAC steps (1/256th of the output voltage swing).

5) Set up the pulse generator to produce short (~50 ns) pulses spaced 10 μs apart. Observe pulse generator output (counter clock input) on the A scope channel, and the DAC output on the B channel (trigger the scope on A). Set up the scope sweep rate so that there is about one clock pulse per division, and set the B vertical sensitivity to maximum or near maximum. This arrangement will allow you to see several "stretched out" segments of each DAC cycle and makes for an interesting display. Now vary the clock pulse width between a 50 ns minimum and just under the clock period. Does the pulse width affect the operation of the 8-bit counter? How does the pulse width affect the DAC output? Explain.

IV. Display the DAC output using the X-Y mode of the scope. Connect the DAC output to the Y input and leave the X input of the scope unconnected. With the scope in X-Y mode, a vertical line should be produced. The length of the vertical line is the DAC output swing divided by the volts/div setting of the Y input. By increasing the Y input sensitivity, you can stretch out the vertical line and clearly see the individual points that comprise it.

1) With the vertical line stretched as much as possible, gradually increase the clock frequency from 100 KHz to the maximum at which the 8-bit counter will function. Explain the "smearing" that this produces. What is the maximum clock frequency that still produces a display not noticeably different from a 100 KHz display?

2) Many oscilloscopes have a TTL-compatible Z input. A +3 volt signal (TTL 1) applied to the Z input of the scope will blank the display. This feature is useful in X-Y mode to prevent unwanted X-Y inputs from being displayed. Design and implement a method of using the scope Z input to allow the maximum frequency observed in part (1) to be increased. What is the new maximum frequency using your method? Although the DAC settling time (300 ns to 1.2 us for MC1408L-8, depending on set-up) limits the maximum frequency, your observed frequency may be less than the maximum set by DAC limitations. Aside from the DAC settling time, explain any problems that limit your maximum frequency, and describe how these problems might be overcome.

PARTS

This project requires a DAC such as the MC1408L-8 and two 4-bit binary counters. Other inexpensive DACs include CMOS devices such as the Analog Devices AD7520 and the Hybrid Systems DAC331.

REFERENCES

Digital-to-analog conversion techniques are described in detail in Barna and Porat [1973], Hnatek [1975], Wait, Huelsman, and Korn [1975], and Tobey, Graeme, and Huelsman [1971].

A10 - ANALOG-TO-DIGITAL CONVERTERS

SUMMARY

Several different techniques for analog-to-digital conversion are discussed in this project. The design of a successive-approximation analog-to-digital converter (ADC) is required.

ADC PRINCIPLES

All ADCs require at least one analog comparator, a device that accepts two analog input voltages and produces a single digital output. As shown in Figure A10-1, the output is 1 if $V_+ \geq V_-$ and it is 0 if $V_+ < V_-$.

The fastest conversion technique, called a *combinational ADC*, uses 2^n-1 comparators to achieve n bits of accuracy. As shown in Figure A10-2, each comparator (C_i) has a reference voltage of $V_{max} \cdot i/2^n$ for the V_- input and the analog input voltage to be converted for the V_+ input. All of the comparators whose references are less than the analog input have a 1 output, and the rest have a 0 output. The comparator outputs are connected to a digital priority encoder, a

FIGURE A10-1 Analog comparator.

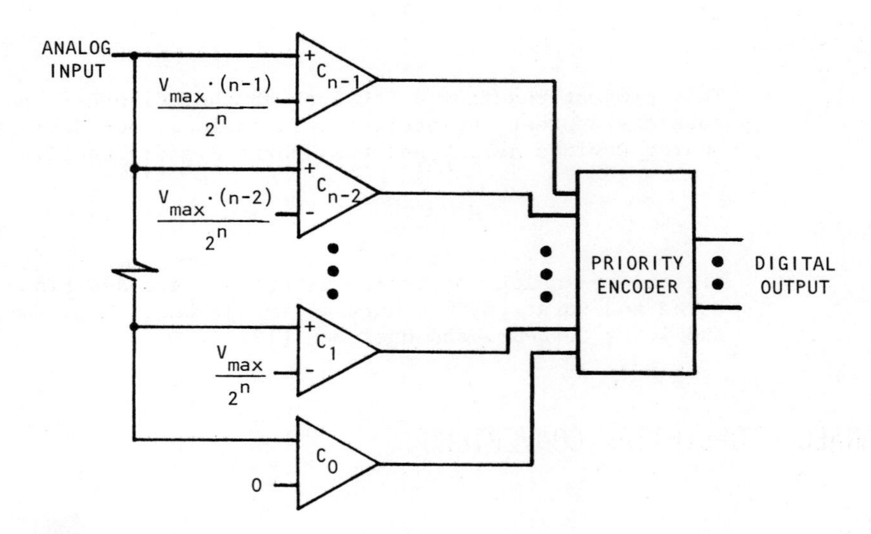

FIGURE A10-2 Combinational ADC.

2^n-input combinational circuit that produces the n-bit encoding of the position of its highest 1 input. This technique is very fast but it is also very expensive.

The most common conversion techniques all have block diagrams similar to Figure A10-3. A result register holds the final converted digital value. The output of the result register is connected to a DAC that produces a reference voltage. The analog input to be converted and the reference voltage are compared, and a control circuit adjusts the value in the result register until the input and reference voltages match. This technique, called a *sequential ADC*, is slower than a combinational ADC because settling time for the digital circuitry, DAC, and comparator must be provided at each of many steps.

A number of different sequential ADC techniques exist, and they differ only in the way that the control circuitry carries out the search for the final result. The most common techniques are described below.

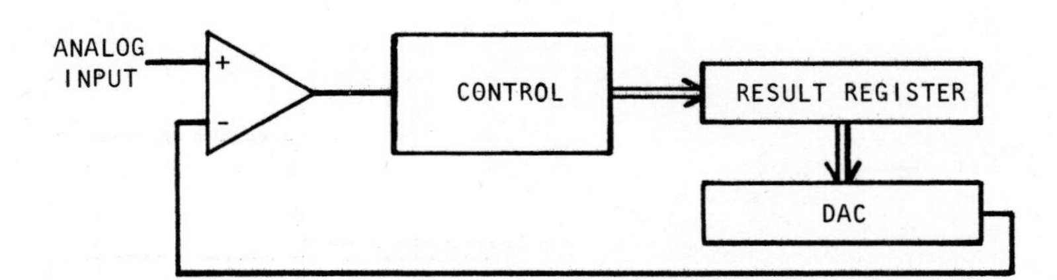

FIGURE A10-3 Sequential ADC.

COUNTER RAMP ADC

A *counter ramp ADC* is the slowest and simplest of sequential ADC techniques. The result register is an n-bit counter that is cleared to begin a conversion. As long as the corresponding DAC output is less than the analog input, the counter is allowed to count up. When the DAC output becomes greater than the analog input the counter is stopped and its stored value is the result. This technique requires a maximum of 2^n steps to perform an n-bit conversion. In this and all sequential techniques the time taken for each step must be greater than the digital control, DAC, and comparator settling times.

SUCCESSIVE-APPROXIMATION ADC

An n-bit *successive-approximation ADC* requires only n steps to perform an n-bit conversion by effectively performing a binary search of the 2^n possible values of the result. The result register is initially set to 10...0, and the comparator indicates whether the input is greater or less than half the full-scale DAC output voltage. Based on this comparison, the high-order bit of the result (bit n-1) is set to either 0 or 1 and remains at this value for the rest of the conversion. In the next step, the next lower bit (bit n-2) is set to 1, and the comparator determines and sets the final value of this bit. The process continues for n steps, until bit 0 is finally tested and set to its final value. At each step in the process the final result becomes known to one more bit of accuracy.

The design of the control circuitry for a successive-approximation ADC is straightforward. The n bits of the result are tested and set one at a time, beginning with bit n-1. Let T_i be a signal that is 1 when bit i is being tested, 0 otherwise. To start a conversion the result register is cleared (or perhaps set to 10...0). At each step, bit i is set to 1 if T_i=1. After the DAC and comparator have settled, bit i is reset to 0 if the analog input is less than the DAC output, otherwise it remains at 1. It is usually possible to set bit i according to the comparator result and set bit i-1 to 1 for the next step simultaneously.

A successive-approximation DAC is the fastest sequential ADC that can be designed for conversion of "random" analog signals.

TRACKING ADC

When an analog signal is known to change only slightly between conversions, a converter known as a *tracking ADC* can be used. The result register of a tracking ADC is an up/down counter. When a conversion is initiated, the analog input and the current DAC output are compared. If the DAC output is less than the analog input then the counter counts up, otherwise it counts down, until the comparator output changes. If the input is changing slowly then only a few counts are needed for the conversion.

Consider what happens if a constant analog input is applied to the tracking ADC described above. On successive conversions the result register will oscillate between two values, since the DAC output will alternately be greater than and less than the analog input. This behavior can be avoided by using a dual-threshold comparator that provides two digital outputs. One output is 1 when $V_+ > V_- + \Delta$ and the other is 1 when $V_- < V_+ - \Delta$. If the inputs are within Δ of each other then both outputs are 0. With such a comparator the digital control circuitry can be designed to update the result register only when the analog input and DAC voltages differ by more than Δ.

ANALOG MULTIPLEXERS

When n different analog inputs are to each be converted to digital form f times per second, it is often economical to provide one fast ADC capable of performing nf conversions per second and an *analog multiplexer* to select among the n different sources. The analog multiplexer acts as a single pole, multiposition switch whose position is controlled by a digital input word.

SAMPLE-AND-HOLD CIRCUITS

If the analog input signal to an ADC may change significantly during the conversion, a circuit called a *sample-and-hold* is needed. The sample-and-hold circuit samples the input voltage for a short period of time preceding the conversion and holds a constant output voltage for the duration of the conversion. The sample-and-hold must be used if the variations of the input voltage during the conversion time exceed the resolution of the ADC.

ASSIGNMENT

Design an 8-bit successive-approximation ADC. You may use an MC1408L-8 DAC and an analog comparator such as the National LM311. A wiring diagram for the LM311 for this application is shown in Figure A10-4. The range of input voltages that can be converted depend on the output range of the DAC and input range of the comparator.

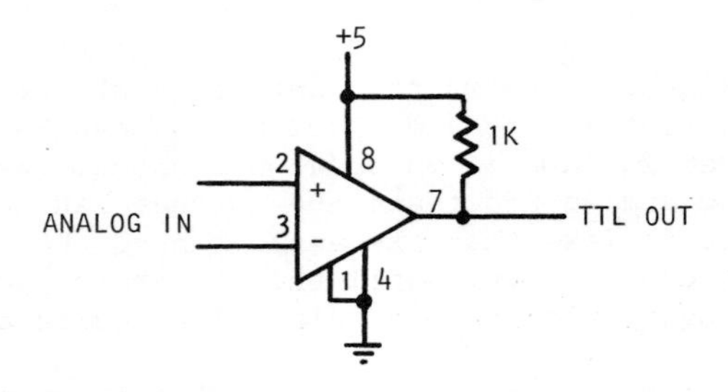

FIGURE A10-4 Wiring diagram for LM311N.

Your circuit should have the following inputs and outputs:

V-IN - The analog voltage to be converted.

INIT - A push button to initialize the state of the circuit for a conversion.

CLOCK - The clock input for the control circuitry. The output should be valid after eight clock cycles.

RESULT - Eight lamps that display the value of the result register.

PARTS

This project requires a DAC such as the MC1408L-8 and an analog comparator such as the LM311. The control circuitry can be implemented using only parts from the standard kit.

HINTS

Use an 8-bit shift register to generate the T_i signals and use J-K flip-flops for the result register.

REFERENCES

Techniques for analog-to-digital conversion are discussed in Blakeslee [1975], Barna and Porat [1973], Wait, et al. [1975], Hnatek [1975], and Tobey et al. [1971]. A successive-approximation register (SAR) and associated control logic in a single IC package is available from a number of manufacturers, including the Advanced Micro Devices 2502 and the National 74C905. Better yet, fairly fast, complete successive-approximation ADCs are becoming available as a single inexpensive CMOS IC such as the Analog Devices AD7570.

A11 - HIGH SPEED LOGIC

SUMMARY

This project demonstrates the static and dynamic electrical characteristics of a high speed ECL gate. It gives exposure to many of the problems of high speed logic.

ASSIGNMENT

I. Build the circuit below (see PARTS section), and record and carefully plot the output voltage versus input voltage for input voltages of 0 to -5.2 volts:

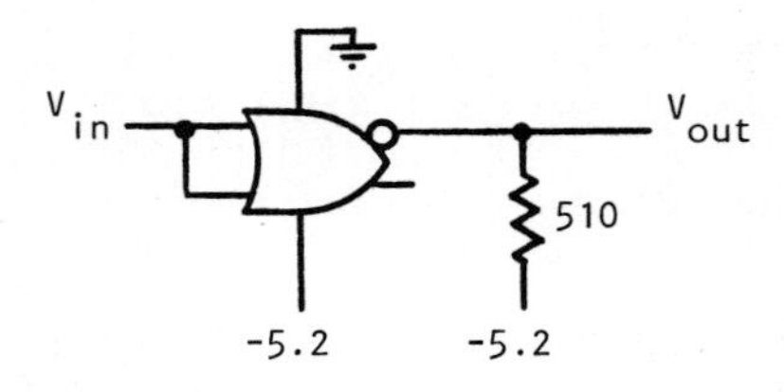

II. Construct the circuit below and use a high speed (≥100 MHz) oscilloscope to measure the propagation delay, rise time, and fall time of the output signal. Then attach a 100 pf. capacitor from the output to ground and re-measure the rise and fall times. Explain the effect of the capacitor (you should examine the output structure of the gate) and why one might be concerned about this effect when designing with high speed ECL circuits.

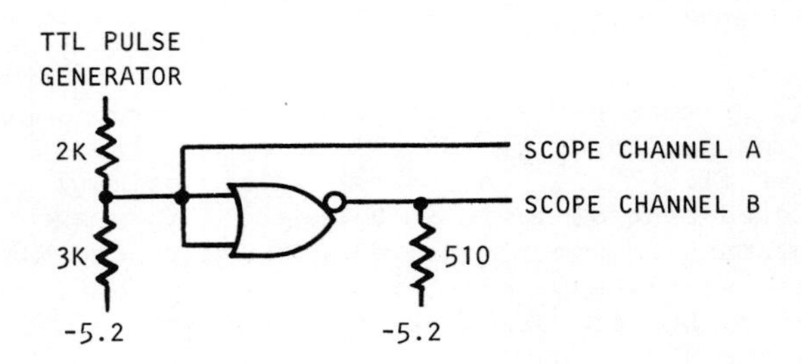

III. Construct the circuit below and sketch the waveform at points A and B. Now place a 100 pf. capacitor from point A to ground and again sketch the waveform at points A and B. What effect did you observe before putting the capacitor on? How did the capacitor change the observed effect, and why? (Consult the circuit diagram for the gate to determine these answers.) Now place the capacitor at point B and again sketch the waveform and explain this effect.

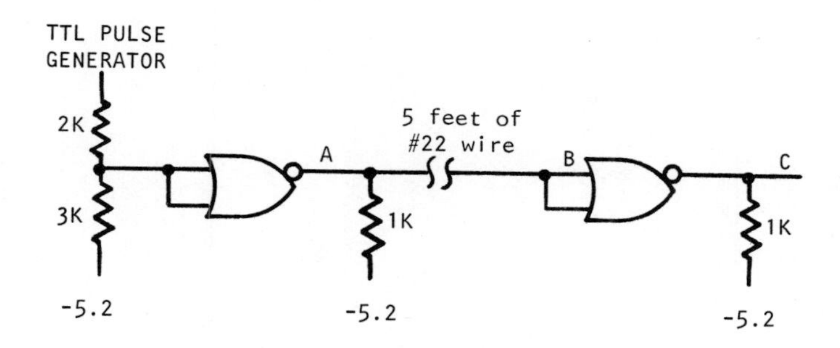

IV. Using the same circuit of Assignment III, sketch the waveform at point C and explain what you see. Explain why one would be concerned with the length of a signal line when designing with ECL. (Similar constraints hold true for TTL, though the restriction is not as tight.)

V. Why are the 1*K* resistors present in the circuit of III and IV? (See what difference they make when they are removed.). How would one prevent the problems encountered in Assignments III and IV if the logic family could drive low impedance loads (say, 50 ohms)? (Consult a text dealing with transmission lines for the answer.)

PARTS

This project can be completed with any 10,000 series ECL gate (10100 through 10112) and the discrete components shown in the figures.

B-Series
5. Two-Week Projects

B1 - CLOCK CONTROLLER

SUMMARY

This project requires the design of a controller to gate a preset number of pulses from a free-running clock. The project provides exposure to pitfalls in synchronization.

ASSIGNMENT

There are often situations where one wishes to pass a predetermined number of clock pulses and then stop, without producing any shortened pulses or glitches. You are to design and construct a circuit that does this.

Your circuit should have eight toggle switches for setting a number *N*, a free-running clock input that can be connected to a pulse generator, and a START push button. There should be a single normally high output Z. When START is pushed, exactly *N* clock pulses should be passed to Z, as shown in Figure B1-1.

Note that the START signal is not synchronized with the clock, and so it can change at any time. In spite of this, no shortened pulses should appear at Z. However, you may assume that the start signal is at least one clock period in length.

Test your circuit by using the Z output to drive another 8-bit counter. The output of the second counter, displayed in eight lamps,

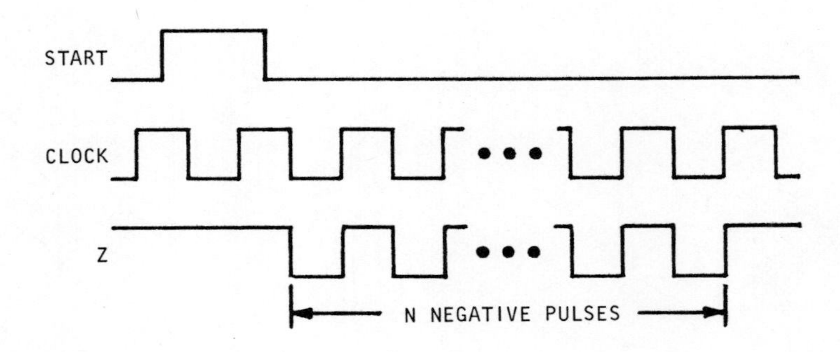

FIGURE B1-1 Clock controller waveforms.

will match the input number *N* if your circuit passes the correct number of pulses. Use a 10 MHz clock (TTL) or a 1 MHz clock (CMOS) for your test.

PARTS

In addition to the parts in the standard kit, you may use two 4-bit programmable binary counters such as 74161, 74163, 74177, 74191, 74193, or 74197. The most convenient counter for this project is the 74191. Alternatively, use two BCD counters such as 74190s.

HINTS

Make full use of the control inputs and outputs available on the 74191 - a clever implementation requires only one or two packages in addition to the 74191s. If you have only up-counters, load the complement of the input number and count up to 11111111 to get the proper number of output pulses. In any case be sure that your circuit functions properly when the input is 00000000.

B2 - LAMP PING-PONG

SUMMARY

The design of a Ping-Pong machine for two human players is carried out in this project. The machine is an amusing application of shift registers as well as an exposure to control logic.

ASSIGNMENT

You are to design and construct a circuit that allows two people to play Ping-Pong using twelve indicator lamps in a row to display the moving ball and using two push buttons for paddles.

Use a 12-bit shift register containing a single 1 to represent the ball. Use one push button for the right-hand user, one for the left-hand user. The ball should be moved by a clock source that causes a shift once every quarter second or so. When the ball is moving to the right, the right-hand player must depress his push button at the time that the ball is in its rightmost position. (The circuit should be sensitive only to the leading edge of the push button signal.) If the player pushes at the correct time, the direction of shift should be reversed and the ball should move toward the left. The left-hand player must then return the volley by pressing his push button when the ball is in its leftmost position. If a player presses his push button too soon or too late, the ball should disappear at the end. When this happens, the player who has just scored should be able to restart the game (serve) by pressing his push button.

PARTS

In addition to parts in the standard kit, you will need three 4-bit bi-directional shift registers, such as 7499 or 74194. Or you may use 4-bit parallel-in, parallel-out shift registers such as 7495, 7496, 74178, 74179, or 74195, externally wired for the opposite shift direction.

OPTIONS

A number of options can be provided to make the came more interesting and to make the project comparable to a C-series project in difficulty.

First of all, provide a number of different rates at which the ball can move. The rate of ball movement should be selected automatically as each player returns a volley. The rate selected should be a function of how the volley is returned, perhaps with the fastest rate selected when the volley is returned at the last possible moment. To implement this feature you will need control circuitry to select a rate, a register to store the rate, and a counter or preferably a BRM to generate the different rates.

A second option is to keep score. Each player's score should be kept in a 4-bit counter. The 4-bit scores should be displayed in the lights whenever the ball is out of play.

B3 - COIN DETECTOR

SUMMARY

This project requires the design of an asynchronous sequential circuit. It provides exposure to pitfalls presented in theory.

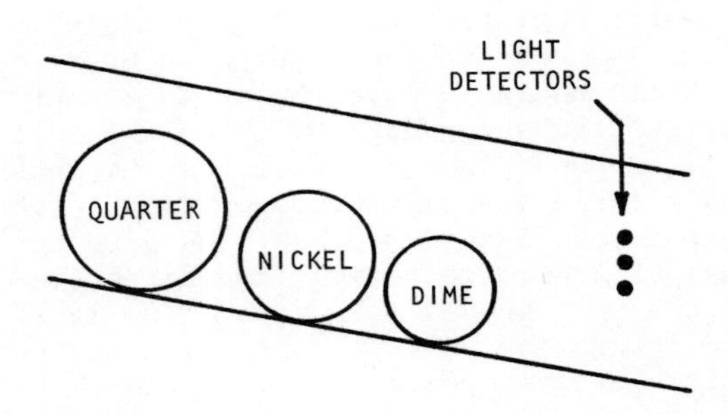

FIGURE B3-1 Coin detector mechanism.

ASSIGNMENT

A device has been designed as part of a coin changer. It is intended to differentiate between quarters, nickels, and dimes by means of their size. It consists of a chute through which a coin rolls, a light source, and three strategically placed light-sensitive diodes, as shown in Figure B3-1.

The light detectors are amplified such that three TTL level outputs are produced. These signals are to be processed by a 3-input 3-output asynchronous circuit that detects the coin's passing and then outputs a 1 on one of three lines (according to the type of coin) until the next coin arrives. The operation of the circuit is summarized in Figure B3-2; IQ, IN, and ID are the three inputs, and Q, N, and D are the outputs.

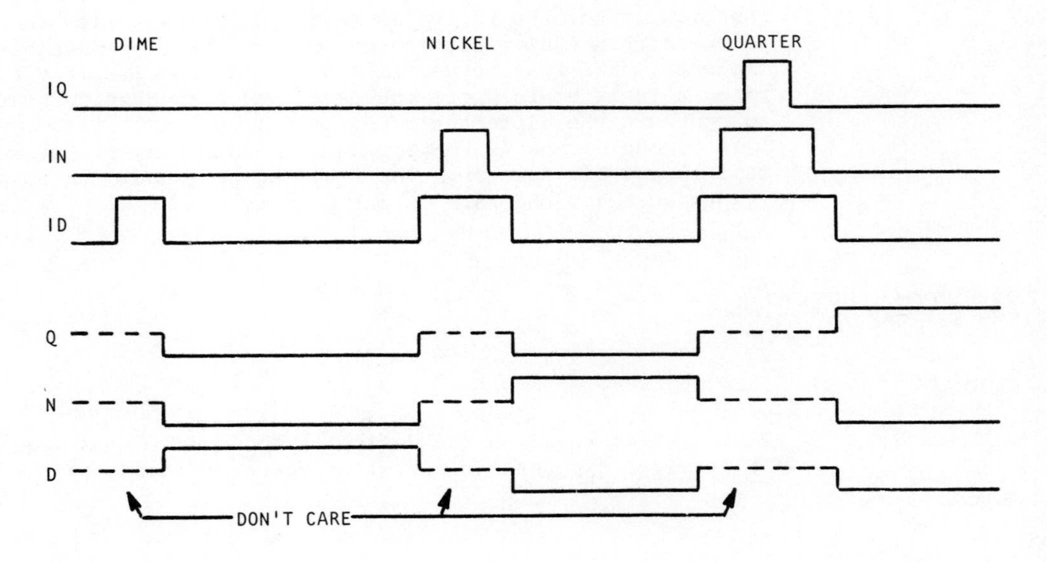

FIGURE B3-2 Coin detector waveform.

Design an asynchronous circuit, using only NAND gates, that will realize the above function. Sketch the actual waveform produced by your circuit in response to the inputs shown in Figure B3-2.

PARTS

Use only NAND gates from the standard kit - no flip-flops.

HINTS

Rather than doing an "intuitive" design, use your skills at formal sequential circuit synthesis, going from word description to primitive fundamental mode flow table to state table to excitation table to final circuit.

B4 - THREE-STATE LOGIC PROBE

SUMMARY

The design of a logic probe for debugging digital circuitry is carried out in this project. The analog circuitry for the probe is given; only the controlling sequential logic must be designed.

LOGIC PROBES

Several commercial firms sell hand-held logic probes that are useful for debugging digital circuitry. The probes have a metal tip that can be touched to a lead in a circuit, and they have a visual output that indicates whether the lead under test is at a logic 1 (high), logic 0 (low), or intermediate level. Detection of the intermediate level is quite useful in debugging since an unconnected ("floating") input gives the intermediate indication.

As shown in Figure B4-1, a logic probe has analog circuitry to translate input voltage levels to some digital encoding of the three possible values: low, high, and intermediate (or floating). Most logic probes also have sequential circuitry to aid the user's detection of short pulses. The low and high output indications of the probe are latched for at least 50 ms even if the input changes after

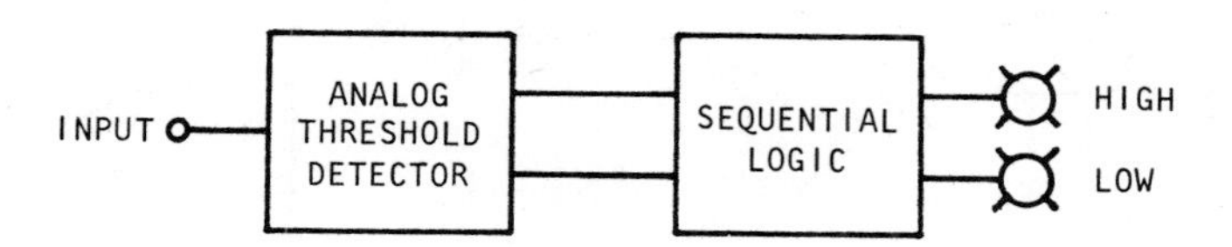

FIGURE B4-1 Block diagram.

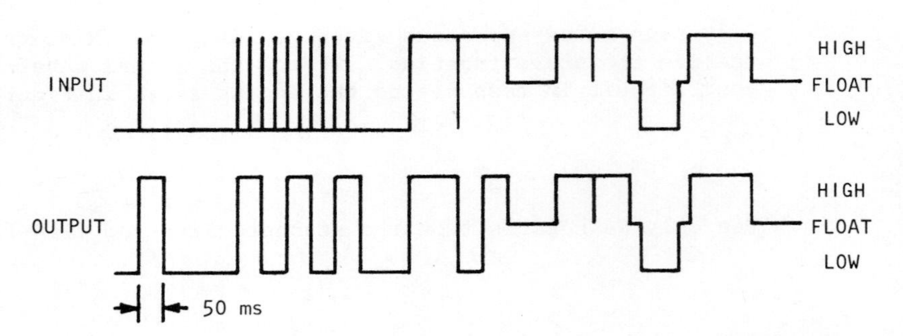

FIGURE B4-2 Logic probe waveform.

a shorter period of time. This is accomplished by starting a timer whenever the displayed output changes to 0 or 1. While the timer is running, the output remains constant but the input is constantly checked for the occurrence of the logic value opposite the displayed output. If that logic value occurs, then the output is changed at the end of the timer period. If the input does not change until after the end of the timer period, then the output changes immediately, that is, it follows the input.

The response of a logic probe to an input signal is shown in Figure B4-2. Short pulses are stretched to give a visual indication of 50 ms, but longer pulses have their true length displayed. The behavior shown in Figure B4-2 indicates that short periods of time at the intermediate level are not stretched. One reason for this is that the intermediate level is normally produced by a constant condition, for example, an open input connection, and pulse detection is not required. Also, slower logic families such as CMOS may spend enough

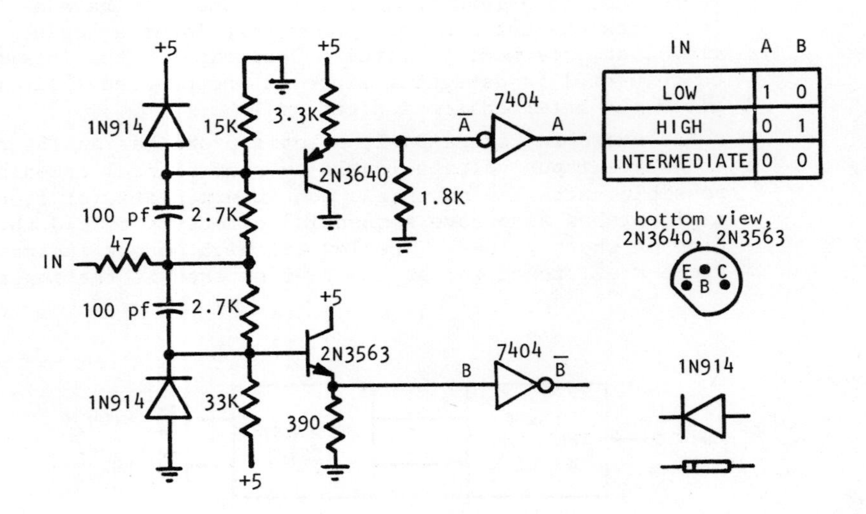

FIGURE B4-3 Analog threshold detector.

time in the intermediate region during transitions to produce unwanted intermediate indications. The waveforms of Figure B4-2 show that the logic probe circuitry gives low and high inputs preference over intermediate inputs.

ANALOG THRESHOLD DETECTOR

Analog circuitry for detecting voltage levels for a logic probe is shown in Figure B4-3. The two diodes provide protection against input levels above V_{CC} or below ground. The top transistor is on and A is 1 when the input voltage is low. The bottom transistor is on and B is 0 when the input voltage is high. Both transistors are off and A and B are 0 when the input voltage is intermediate. The resistor network is designed so that an intermediate voltage is simulated when the input is floating. The transistors shown are faster than a standard TTL gate, and so the circuit is capable of following any input pulses that can be detected by TTL.

ASSIGNMENT

Design a logic probe that uses the analog threshold detector of Figure B4-3 and sequential circuitry to stretch pulses as indicated in Figure B4-2. The output of the probe should be two lamps, HIGH and LOW, that show the logic level present. For floating inputs both lamps should be off. Test your circuit by connecting its input to a pulse generator, and verifying that it catches both short low pulses and short high pulses. Verify that for low frequency inputs (e.g., square waves < 10 Hz) the probe output follows the input exactly. Answer the following questions:

1) What are the low and high thresholds of the level detector?

2) What is the length of the shortest high pulses that your circuit will detect? low pulses?

3) What determines the minimum length of detectable pulses? How can this length be decreased?

PARTS

In addition to parts in the standard kit, you will need the parts shown in Figure B4-3 and a one-shot or timer such as a 74121, 74122, 74123, 9602, or 555 to provide the 50 ms delay.

HINTS

Divide your circuitry into two parts: one section that latches up an input pulses when the opposite value is being displayed, and another part that displays the latched pulses and resets the input latches at the end of the timer period. Keep the wiring in the analog circuit short.

OPTIONS

Instead of the two-lamp output format, use a 7-segment display. The display should read L, H, or F depending on the output state. Is this output format more convenient for a user of the probe?

Another option is to use a free-running 20 Hz clock rather than a 50 ms delay to determine when the probe output may change. In this scheme, the latched inputs would be transferred to the probe output at, say, a rising edge of the clock, and the output would remain constant until the next rising edge. The output of this probe for short and very long input pulses would be effectively the same as the previous designs, but for signals in the 5-20 Hz range the output can be noticeably different. Explain the reason for and the effects of this difference.

B5 - T-BIRD TAILLIGHTS

SUMMARY

The design of a circuit that emulates the operation of the taillights of a Thunderbird is carried out in this project.

ASSIGNMENT

You are to design and build a circuit that controls the six taillights on a Thunderbird. Use six lamps to simulate the six taillights (three on each side of the car), and use two toggle switches for the turn signals. One toggle switch should be used to indicate a left turn, one toggle switch for a right turn. If both toggle switches are set, then the emergency flasher should be activated, since the driver obviously doesn't know what he is doing.

For a right turn, the three right-hand lamps should be activated and the left-hand lamps should be off. The three lamps should cycle on and off as shown in Figure B5-1, and one cycle should take about one second. Operation of the left-hand lamps for a left turn is analogous. When the emergency flasher circuit is enabled, all six taillamps should flash on and off in unison, with a frequency of about 1 Hz.

Your circuit should also emulate brake lights. Use a push button to simulate the brake pedal. When the brakes are on, all six taillights should be on continuously, except if the right or left turn signal is on. In that case the three taillights for the turn signal should operate normally, and the other three should be on continuously.

● ● ● → ○ ● ● → ○ ○ ● → ○ ○ ○ → ● ● ●

● OFF
○ ON

FIGURE B5-1 Taillight sequence.

Also emulate parking lights. Use a toggle switch to simulate the parking light switch. When the parking lights are on, the taillights should be at half brightness at all times when they would otherwise be off.

PARTS

This project can be completed using only parts from the standard kit.

HINTS

Note that the brakes override the emergency flasher.

For the parking lights connect a 60 Hz or faster clock signal to the lamps to produce half brightness. Varying the duty cycle (fraction of "on" time) varies the lamp brightness.

OPTIONS

Buy a new Thunderbird and install your circuit in it.

B6 - "ADD-AND-SHIFT" BINARY MULTIPLIER

SUMMARY

This project requires the design of a binary multiplier for two 4-bit operands using the conventional "add-and-shift" algorithm.

ASSIGNMENT

Design a binary multiplier for two 4-bit unsigned operands using the conventional "add-and-shift" algorithm. The circuit should produce an 8-bit product after four steps, as shown below for the example multiplication of 0101 by 1101.

0101	5	Step 0:	00000000
×1101	×13	Step 1:	+0101
01000001	65		01010000
			→00101000
		Step 2:	+0000
			00101000
			→00010100
		Step 3:	+0101
			01100100
			→00110010
		Step 4:	+0101
			10000010
			→01000001

Your circuit should have eight toggle switches for the two input operands and eight lamps for the output. There should be a single push button for performing a multiplication. When the button is pushed, the product register should be cleared, and when it is released the four cycles for executing the multiplication should take place automatically.

PARTS

This project can be completed using only parts in the standard kit.

C-Series
6. Two-Week Projects

C1 - SERIAL ARITHMETIC AND LOGIC UNIT

SUMMARY

The design of circuits for performing various operations on serial data is carried out in this project. The project involves combinational logic design and the use of shift registers.

ARITHMETIC AND LOGIC UNITS

An *arithmetic and logic unit (ALU)* is a circuit that performs various arithmetic and logical operations on two input operands. Typical ALUs perform binary operations such as add, subtract, AND, OR, XOR; unary operations such as increment, decrement, transfer input directly to output; and constant operations such as output constant 0 or 1. The 74181, 74281, and 74381 are examples of MSI ALUs that operate on 4-bit parallel data. It is also possible to design serial ALUs for which the two input operands and the output are stored in shift registers and processed one bit at a time. The advantage of a serial ALU is that only one copy of the circuitry for one ALU bit is needed to process n-bit data, rather than n copies for a parallel ALU; the disadvantage is that n clock cycles are required to process the data.

ASSIGNMENT

I. Design a 6-input, 2-output combinational ALU circuit that processes two 1-bit operands. The circuit should have two data inputs A and B, a carry input CI, three function select inputs F1,F2,F3, a data output S, and a carry output CO. The circuit should perform eight functions as shown in Table C1-1. For each operation the value of S should be computed as shown; for the arithmetic operations (increment, negate, add, subtract) the value of CO should be computed as the carry from the indicated addition.

TABLE C1-1 ALU Functions for Assignment I

F_1 F_2 F_3	Operation	Name
0 0 0	$S \leftarrow 0$	Clear
0 0 1	$S \leftarrow \overline{A}$	Complement
0 1 0	$S \leftarrow$ A plus CI	Increment
0 1 1	$S \leftarrow \overline{A}$ plus CI	Negate
1 0 0	$S \leftarrow B$	Transfer B
1 0 1	$S \leftarrow A \oplus B$	Exclusive OR
1 1 0	$S \leftarrow$ A plus B plus CI	Add
1 1 1	$S \leftarrow$ A plus $\overline{B}$ plus CI	Subtract

II. The ALU circuit of Assignment I can be used to serially process operands of any length. Complete the design of the system shown in Figure C1-1, which uses two shift registers to hold 4-bit operands A and B. The system should have the following inputs and outputs:

F1,F2,F3 - These three toggle switches are used to set the desired function.

B-IN - These four toggle switches determine the value of the B-register set by a parallel load.

LOAD - This push button is used to load B-IN into the B-register and also should set the carry flip-flop with the proper initial carry for arithmetic operations, as determined by F1,F2,F3.

CLOCK - This push button controls the shifting of A and B and loading of the carry flip-flop for performing operations. The result should be contained in A, and B should be restored to its loaded value, after four clock pulses.

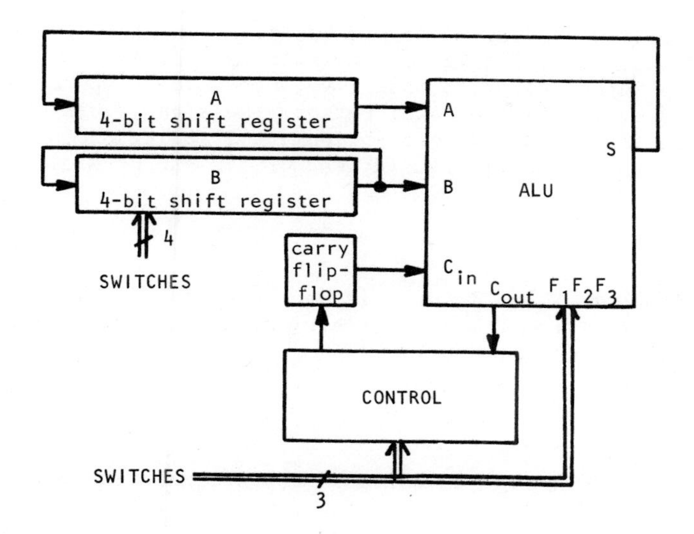

FIGURE C1-1 Block diagram.

A-REG - These four lamps should show the contents of the A-register.

B-REG - These four lamps should show the contents of the B-register.

C - This lamp should show the contents of the carry flip-flop.

Test your circuit by performing each operation with a number of different operands.

TABLE C1-2 ALU Functions for Assignment III

Operation	Name
S ← A plus CI	Increment
S ← A plus B plus CI	Add
S ← A plus $\overline{B}$ plus CI	Subtract
S ← 0	Clear
S ← 1	Preset
S ← $\overline{A}$	Complement
S ← B	Transfer B
S ← A ⊕ B	Exclusive OR

III. The 74181 4-bit ALU has five function selection inputs, M,S0,S1,S2,S3. Design a combinational translation circuit that maps three inputs F1,F2,F3 into five outputs so that three switches can be used to make the 74181 perform the eight functions shown in Table C1-2. Then repeat Assignment II using one bit of a 74181 and the translation circuit instead of the ALU of Assignment I.

PARTS

Assignments I and II of this project require only integrated circuits from the standard kit. For Assignment III you will need in addition a 74181 4-bit ALU.

HINTS

In Assignment III you can make any assignment you wish of the functions of Table C1-2 to the eight values of F1,F2,F3. Also, for some of the functions, you have the choice of a number of different ways to perform that function using the 74181. Try to make these choices in a way that minimizes the translation circuitry between F1,F2,F3 and M,S0,S1,S2,S3.

C2 - FLOATING-POINT FREQUENCY COUNTER

SUMMARY

This project requires the design of a circuit that measures input frequencies up to 10 MHz and displays the measured frequency with two-digit accuracy.

FREQUENCY COUNTERS

The basic structure of a frequency counter is shown in Figure C2-1. A measurement is begun by clearing the BCD counter and the reference source. Then the reference source enables the counter to count the input pulses for a precise interval, say, 1, 0.1, or 0.01 sec. The input frequency is proportional to the number of pulses counted during the reference period. The reference period is chosen so that the constant of proportionality is a power of 10, and only a shift of the decimal point is needed for scaling.

In this project, the reference period will be generated by a one shot. However, in commercial counters a much more accurate method is used. An accurate crystal oscillator is used to generate a precise high frequency signal, say, 10 MHz. This signal is then "divided down" by a counter chain to produce a reference period with the same precision as the crystal oscillator.

The resolution of a frequency counter is determined by the length of the reference period and the number of digits that are displayed.

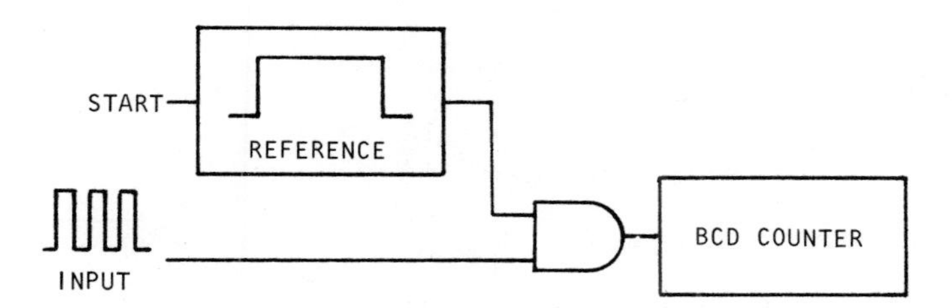

FIGURE C2-1 Block diagram.

The accuracy depends mainly on the accuracy of the reference period. There is also an uncertinty of ±1 count because the enable gate and the input frequency are not synchronized. This effect is only significant at very low frequencies. At low frequencies it is better to use a method that measures the period and inverts it to find the frequency.

ASSIGNMENT

You are to design a frequency counter that can count up to 999,999 input pulses during a reference period of 1 second, so that frequencies up to 1 MHz can be measured. It is desired to display only the two most significant digits of the measured frequency and a scale factor. One way this could be done is to measure the frequency exactly and scale the result at the end of the measurement. A different approach, to be used here, is to design a special frequency counter that counts in "floating-point."

Your counter should have three digits of output, E, D_0, D_1, to represent the number $(10D_1 + D_0)10^E$. Initially two BCD counters for D_1 and D_0 are cleared and E is set to 0. The number of input pulses during a 1 second reference period should be counted. Initially the input signal should be fed directly to the 2-digit counter D_1D_0. If the count reaches 99 then the represented frequency is 99×10^0. The next count should change the output to 10×10^1 and should connect the D_1D_0 counter to the input frequency divided by 10. Hence D_1D_0 will now be incremented once for every 10 input pulses, as it should since the output is now $D_1D_0 \times 10^1$. Likewise when the output reaches 99×10^1, the next D_1D_0 count should change the output to 10×10^2 and connect D_1D_0 to the input frequency divided by 100. You should provide sufficient circuitry to continue this process for displaying a maximum count of 99×10^4.

Your circuit should have an input for the signal to be measured and a single push button for performing a measurement. Provide a 1 second reference signal and test your system by measuring various frequencies up to 1 MHz. Then reduce the reference signal to 0.1 second. With standard TTL counters your system should be capable of measuring frequencies up to 10 MHz.

PARTS

In addition to parts in the standard kit, you will need the following:

1 - BCD counter with parallel load. Suggested: 74160 or 74162. Alternates: 74190, 74192

5 - Any BCD counter.

1 - Timing monostable. Suggested: 555, 9602, 74121, 74122, 74123.

HINTS

Use a counter for E and use E as the input of a multiplexer to select the input frequency or submultiple to be connected to the D_1D_0 counter.

OPTIONS

When using a 0.1 second reference period, provide circuitry to initialize E at 1, not 0, so that the true frequency is displayed.

C3 - SERIAL BINARY MULTIPLIER

SUMMARY

This project requires the design of a 4-bit multiplier using serial techniques. It provides exposure to shift registers and control logic design.

ASSIGNMENT

Design a serial multiplier to form the 8-bit product of two 4-bit operands. Use an "add-and-shift" algorithm in which the individual 4-bit additions are performed serially. Your circuit should have the following inputs and outputs.

MULTIPLIER - These four toggle switches are used to set the value of the multiplier.

MULTIPLICAND - These four toggle switches are used to set the value of the multiplicand.

INIT - This push button initializes the multiplier for a new multiplication. Pushing this button should clear the product register, load the operands set in the switches into two shift registers, and initialize the control circuitry to the proper state.

CLOCK - This push button is used to actually perform the serial multiplication. A worst-case multiplication should require about 20 clock pulses.

DONE - This lamp, reset by INIT, should turn on when the multiplication is completed.

PRODUCT - These 8 lamps display the product.

You should provide circuitry to shorten the multiplication time when there are zeroes in the multiplier. Briefly indicate how your circuit could be extended to multiply 16-bit operands.

PARTS

You may use one 4-bit shift register in addition to the two shift registers and other parts in the standard kit.

HINTS

The final 8-bit product can be held in a 4-bit accumulator register and the original 4-bit multiplier register. As the multiplication progresses, bits of the multiplier are discarded, making room for the expanding product.

OPTIONS

Provide extra control circuitry and a free-running clock so that a multiplication proceeds automatically.

C4 - BINARY/BCD CONVERSION

SUMMARY

This project requires the design of circuits for serial binary-to-BCD conversion and parallel BCD-to-binary conversion. It provides exposure to intuitive design techniques using MSI.

ASSIGNMENT

I. Design a circuit that serially converts a 7-bit unsigned binary number into its 3-digit BCD representation. Your circuit should have seven toggle switches for specifying the binary number and a push button for loading the number into a shift register. There should also be a clock push button for serially shifting the input number into the conversion circuits; after seven clock pulses the conversion should be complete. The output consists of three 4-bit BCD digits (but note that for this project the largest possible value of the most significant digit is 1). Use 7-segment displays, if available, to display the BCD output. Your circuit should process the most significant bit first, as shown in Table C4-1 for an example conversion, $1011011_2 \rightarrow 91_{10}$.

TABLE C4-1 Binary-to-BCD Conversion

BCD		Binary
000		1011011
001	←	0110110
002	←	1101100
005	←	1011000
011	←	0110000
022	←	1100000
045	←	1000000
091	←	0000000

II. Design a combinational circuit that converts a 3-digit BCD number less than 128 intc the 7-bit binary representation. The circuit should have nine toggle switches to enter the BCD number (two 4-bit digits and one bit for the most significant digit). The output should be displayed with 7 lamps.

III. Connect the output of the binary-to-BCD converter to the input of the BCD-to-binary converter. Test the composite circuit by loading numbers into the binary-to-BCD converter and observing the output of the BCD-to-binary converter as conversion is executed.

PARTS

In addition to parts in the standard kit, for Assignment I you will need the following:

1 - 8-bit parallel-in shift register for holding the input number. Recommended: 1 × 74166 or 2 × any 4-bit parallel-in shift register.

1 - 7483 adder (in addition to the one in the standard kit).

2 - 7-segment displays and drivers.

1 - ±1 or 7-segment display and driver.

Assignment II can be completed using only three 7483 adders.

HINTS

In Assignment I, design a circuit that processes one BCD digit, so that *n* of these circuits can be cascaded to perform an *n*-digit conversion. For the conversion from 7 bits to 3 digits you will really need only two of these circuits, because the most significant digit takes on only two of the ten possible values - 0 and 1. The circuit to process one digit consists of a 4-bit shift register and circuitry to correct its counters and inform the next stage when its contents exceed 9.

As mentioned earlier, Assignment II can be completed using only three 7483 adders and no other components.

OPTIONS

Modify the circuit of Assignment I to convert signed numbers (7 bits plus sign) in the ones' complement representation to 3 digits plus sign. What problems occur in serially converting two's complement numbers to BCD?

C5 - UNIVERSAL SEQUENTIAL MACHINE

SUMMARY

A sequential machine can be simulated if its state/output table is stored in memory. This project requires the design of a system for simulating small sequential machines.

ASSIGNMENT

The state/output table of a 2-input, 8-state asynchronous sequential machine has 8 rows, one for each state, and 4 columns, one for each input combination. Hence, there are a total of 32 entries. Each entry indicates the next state and the output. If there is only one output bit, the information in each entry in the state table can be encoded in 4 bits, one for the output Z and three for an encoding of the state. The entire state/output can be encoded in $4 \times 32 = 128$ bits.

You are to design a system that can store a representation of any such state/output table in a small memory and simulate the behavior of the machine

A block diagram of the system is shown in Figure C5-1. A RUN/PROGRAM switch determines into the memory at the address specified by the I1, I2, y1, y2, and y3 inputs. In the RUN position, the current Y1, Y2, and Y3 memory outputs and the two inputs, I1 and I2, are used to form a new address for the memory. The new address is gated into the address register and the operation is repeated once for each input clock pulse. The clock input should be connected to a pulse generator. The clock rate determines how fast the simulator can respond to a change in I1 or I2; since these signals are controlled by switches,

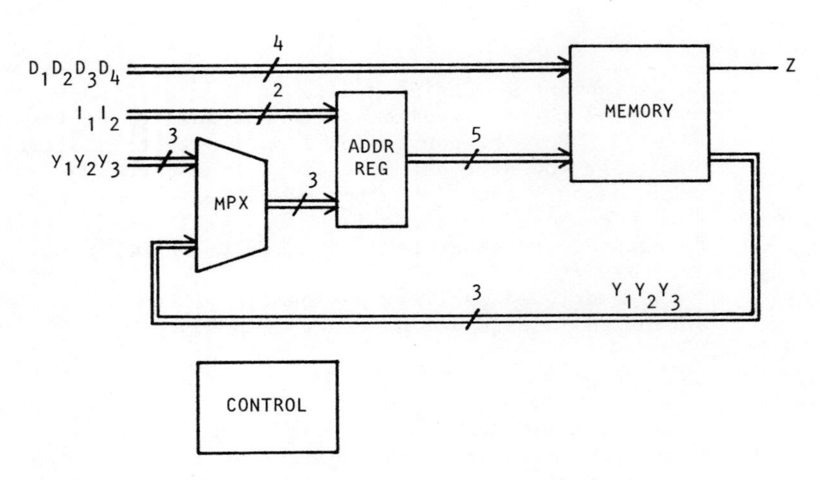

FIGURE C5-1 Block diagram.

60 Hz is sufficiently fast. When the system reaches a stable state in the simulated machine, the operation of the clock continues, but the new address is always the same and the output is stable.

Your machine should have lights for the memory outputs and for the address register outputs, and the following inputs:

RUN/PROGRAM - This toggle switch is described above.

LOAD ADDRESS - In program mode this push button loads the address register with the values of toggle switches I1,I2,y1,y2,y3.

INCREMENT ADDRESS - In program mode, this push button increments the address register.

WRITE DATA - In program mode, this push button causes the data inputs D1,D2,D3,D4 to be written at the current address in the address register.

D1,D2,D3,D4 - These are four toggle switches for data.

I1,I2,y1,y2,y3 - These five toggle switches are used to set the address. The same physical switches can be used for both data and address since they are used at different times.

Note that the toggle switches I1 and I2 should be debounced since they are used when the system is in RUN mode. The other switches should have no effect in RUN mode (except RUN/PROGRAM, of course).

Construct the system and first test the ability to correctly read and write every memory location. Then test the system by deriving, entering, and testing state tables for the following sequential machines:

a) An SR flip-flop (I1 = S, I2 = R, Z = Q)

b) An edge-triggered D flip-flop (I1 = D, I2 = clock, Z = Q)

c) A modulo-4 up/down counter (I1 = up/down, I2 = clock, Z = 1 for states 1 and 2, Z = 0 for states 0 and 3)

d) A machine of your own fancy

The system may be used to simulate a synchronous sequential machine if the pulse generator is disconnected from the system clock input. Instead, the clock signal of the simulated machine is used to clock the system. Using this mode of operation, derive, enter, and test the following sequential machines:

e) An edge-triggered D flip-flop (I1 = D, I2 = don't care, Z = Q)

f) An edge-triggered J-K flip-flop (I1 = J, I2 = K, Z = Q)

Turn in a state/output table for each of the simulated machines.

PARTS

In addition to parts in the standard kit, you will need the following:

1 - 32-word by 4-bit memory. Suggested: 2 × 7489.

1 - 5-bit address register and counter. Suggested: 2 × 74161 or 74163. Alternate: any 5-bit edge-triggered register and 5-bit counter.

HINTS

For proper operation, the loading of the address register in RUN mode must be synchronous, even if the state table being simulated has no races. Why?

Be sure to use the correct state table for an S-R flip-flop, that is, it should model the circuit shown in Figure A5-1(b).

C6 - BINARY DIVIDER

SUMMARY

The design of a fixed-point binary divider is carried out in this project. The divider accepts as inputs an 8-bit dividend and a 4-bit divisor and produces a 4-bit quotient and a 4-bit remainder.

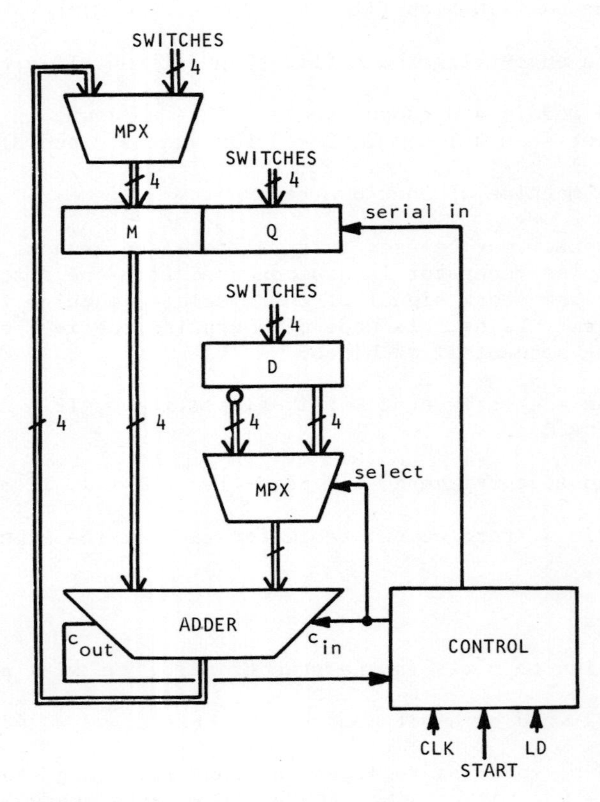

FIGURE C6-1 Block diagram.

ASSIGNMENT

In this project you are to design a divider circuit for unsigned binary integers using the nonrestoring division algorithm. A block diagram for the circuit is shown in Figure C6-1. Two 4-bit shift registers M and Q hold the initial dividend, and a 4-bit register D holds the divisor. At each step of the division the 8-bit MQ register is shifted left and a new quotient bit is shifted in. At the end of the division, Q should contain the 4-bit quotient and M should contain the 4-bit remainder. Your circuit should have the following inputs and outputs.

DIVIDEND - Eight toggle switches to set the value of the dividend.

DIVISOR - Four toggle switches to set the value of the divisor.

LOAD - A push button to load MQ with DIVIDEND and D with DIVISOR. Two push buttons may be used if DIVIDEND and DIVISOR share the same toggle switches.

START - A push button to start a division after the operands have been loaded. The algorithm should proceed automatically (an external free-running clock will be needed).

MQ - Eight lamps that display the M and Q registers at all times.

OVERFLOW - A lamp that indicates when a product greater than 4 bits would be produced, that is, when $M \geq D$. Execution of the algorithm should be automatically inhibited when overflow is present. That is, pushing START should cause the system to halt with OVERFLOW on and MQ unchanged.

PARTS

In addition to parts from the standard kit, this project requires a 4-bit register such as 74175, two 4-bit parallel load shift registers, and a 4-bit, 2-input multiplexer in addition to the one in the standard kit.

HINTS

Before designing and building your circuit, draw a flowchart of your division algorithm and test it with several different operands and exceptional conditions.

REFERENCES

Restoring and nonrestoring division algorithms are discussed in several computer design texts, including Gschwind and McCluskey [1975] and Bartee [1972].

C7 - DIGITAL TACHOMETER

SUMMARY

The purpose of this project is to design and build a digital tachometer suitable for use in an automobile. This is an example of a design where the relatively long times involved allow a slow implementation to be chosen over a more expensive faster one.

TACHOMETERS

A tachometer operates by measuring the frequency of firing of the engine's points. There are two approaches for doing this. The first is to simply count the number of firings in a fixed period of time, as was done in Project C2. Although this is by far the simplest of the methods, it will be shown that this approach is not satisfactory.

The other method is to measure the period between successive firings of the points and then to find the inverse of this, that is, the frequency. This approach can be implemented with a reasonable amount of circuitry by keeping in mind that the time required to find the inverse is not critical.

COMPARISON OF ALTERNATIVES

In comparing the different approaches, it is important to note that the reaction time of the human being is critical, as well as how fast the engine can change RPM (revolutions per minute). The eye can detect very short changes and interpret them in about 0.1 to 0.2 seconds. For this reason, the tachometer's RPM display should be updated within this period of time. The engine can change about 3000 RPM/sec. For the driver to get useful information on the engine speed would require readings accurate to 100 RPM about every .05 seconds or better. The operating range of the display should be about 600 to 7000 RPM with accuracy of about ±50 RPM throughout the range. The points of a typical 4-cylinder or 8-cylinder engine fire four times per revolution. Hence the RPM can be related to the number of point firing pulses per second by the formula:

$$1 \text{ RPM} = \frac{1 \text{ revolution}}{\text{minute}} \times \frac{1 \text{ minute}}{60 \text{ seconds}} \times \frac{4 \text{ pulses}}{\text{revolution}} = \frac{1 \text{ pulse}}{15 \text{ seconds}}$$

$$\text{RPM} = 15 \times (\text{pulses/second})$$

If we take the first alternative of counting the number of pulses within a fixed period of time, it would be most convenient to use the relationship:

$$\text{RPM} = 100 \times (\text{pulses/.15 second})$$

That is, we could count the number of pulses in a .15 second period and append two zeroes to get RPM; the display could be updated every .15 seconds. However, in addition to the fact that the display is not updated quite as fast as we would like it to be, this method has a primary disadvantage of poor accuracy. A difference of only one pulse per .15 second period makes a difference of 100 RPM in the display. If we counted pulses over an even shorter period for faster updating, the accuracy would be correspondingly poorer.

The second alternative is much more accurate: measure the period of the pulses and divide to find the RPM. At 600 RPM the time between pulses is .025 seconds, and at 7000 RPM the period is .0021 seconds. At the low end the period is still short enough that updating the display once per period gives the driver his information quickly enough, and at the high end the period is not so short that the division cannot be carried out in one period by standard TTL or CMOS circuits. The RPM can be given by the relationship:

$$\text{RPM} = 15 \times \text{pulses/second} = \frac{15}{\text{seconds/pulse}}$$

Choosing then the second alternative as the method for designing the tachometer, the first problem we are faced with is how to measure the period between pulses. The simplest method is to count the number of pulses of a high frequency (in relation to 7000 rpm) free-running clock between engine point pulses. The period of the point pulses is then simply the number of clock pulses times the period of the free-running clock. In order to achieve accuracy of ±50 RPM throughout the range of the tachometer, it can be shown that the frequency of the free-running clock should be about 80 KHz or higher (show this as part of the assignment). A choice of 109.2 KHz for the free-running clock frequency leads to the relationship:

$$\text{RPM (in hundreds)} = \frac{15 \times 109.2 \text{ KHz}}{\text{clock pulses/engine pulses}} \times \frac{1}{100}$$

$$= \frac{2^{14}}{\text{clock pulses/engine pulses}}$$

The only problem remaining, then, is how to carry out the division called for above. Shift and subtract algorithms are certainly fast enough to carry out the division within .0021 seconds, but they require too much circuitry and their higher speed is not necessary. A brute force method using counters requires less circuitry. As shown in the block diagram of Figure C7-1, the engine point pulses are fed to the control unit to produce a gating signal for the period counter, which counts the period between pulses in units of the 109.2 KHz clock period. After the period has been counted, a 14-bit counter and a two-digit display counter are cleared, and the contents of the period counter are transferred to a down counter. The down counter counts down and the up counter counts up until the down counter reaches zero. When this happens the display counter is incremented and the down counter is reloaded from the period counter. The process is repeated

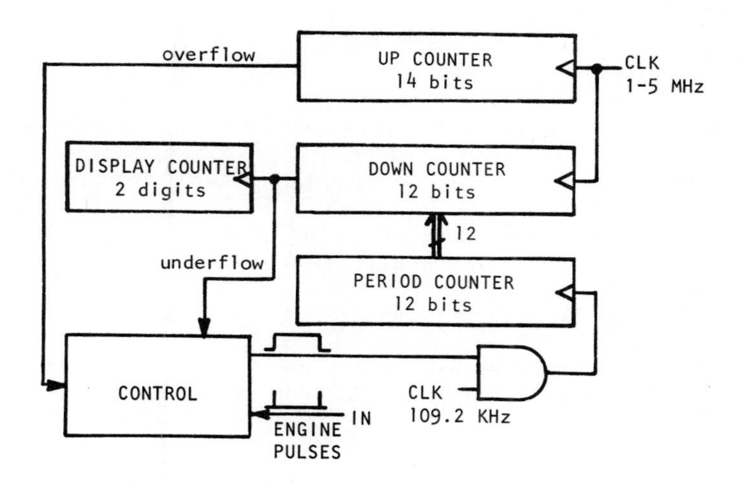

FIGURE C7-1 Block diagram.

until the up counter overflows. At this point the display counter contains the (truncated) quotient of 2^{14} divided by the contents of the period counter. The display counter contains the RPM in hundreds.

ASSIGNMENT

You are to design and build a digital tachometer following the block diagram of Figure C7-1. The tachometer should accept as input a pulse train that simulates the point firings of an engine that makes four firings per revolution. The output of your circuit should be two BCD digits giving the engine speed in RPM. The system should be accurate to ±50 RPM over a range of at least 600 to 7000 RPM.

Referring to the block diagram, note that 12 bits are sufficient for the period and down counters, since the longest period, corresponding to 600 RPM, produces a period count of 2730. Without "pipelining" it is not possible to update the display counter after each point firing. Instead, the control unit should initiate a period count only after the previous period count has been divided into 10,000 and displayed. The clock for performing the division should be a separate fast clock (1 to 5 MHz range). With a 1 MHz division clock, at what rate will the display be updated when the engine speed is 600 RPM? 7000 RPM?

The output digits may be displayed by eight lamps or two seven-segment displays. In any case, the output digits will have to be buffered by latches unless the division time is very short - otherwise the display will flicker whenever a division is taking place.

You will need three separate clock sources for this project - a 109.2 KHz source for measuring the engine pulse periods, a clock in the 1 to 5 MHz range for the division circuitry, and a variable 40 to 500 Hz clock for simulating the engine pulses.

Calibrate your system and test it for various static RPM inputs and for varying RPM inputs that simulate acceleration and deceleration. Comment on the advantages and disadvantages of digital RPM calculation and display for real automobiles.

PARTS

In addition to parts in the standard kit, you will need the following:

1 - 12-bit period counter. Suggested: 3 × any 4-bit binary counter.

1 - 12-bit down-counter. Suggested 3 × 74191 or 74193. Alternate: 3 × any 4-bit parallel-load up-counter and inverters.

1 - 14-bit counter. Suggested: 4 × any 4-bit binary counter.

1 - 2-digit display counter and latch. Suggested: 2 × 74143 or 74144. Alternate: 2 × any BCD counter and 2 × any 4-bit register.

2 - seven-segment displays (optional). Suggested: Any LED or incandescent displays, drivers included in 74143 or 74144 above. Alternate: Any displays and appropriate drivers.

OPTIONS

The same counter can be used for both the period counter and the down counter if a separate latch is provided to store the period during the division process. In this case, a 12-bit register can be substituted for a 12-bit counter.

Another option is to complete the option for Project B5, and then install your tachometer.

D-Series
7. Three-Week Projects

D1 - A SIMPLE PROCESSOR

SUMMARY

A simple processor for 4-bit operands is designed in this project. The processor has four 4-bit registers, a number of 8-bit instructions, and memory for a 16-instruction program. A sample program is a multiplication routine.

FUNCTIONAL DESCRIPTION

This section gives the description of a small processor. Figure D1-1 is a block diagram of the processor as it would appear to a programmer. The processor has all the rudimentary elements of a computer, but it has insufficient features to serve in any practical capacity. The machine has three classes of instructions: register instructions, branch instructions, and halt-I/O instructions, as shown in Figure D1-2.

Register instructions perform an operation on the two registers specified by S and D. The notation (S) denotes the contents of the register specified by S. A function specified by OP and CC is performed on (S) and (D) and the result is stored in (D). OP specifies one of the operations and CC controls the carry-in bit used in the operation. The meanings of OP and CC are detailed in Tables D1-1 and D1-2. The value of C_I for each operation is determined by CC as

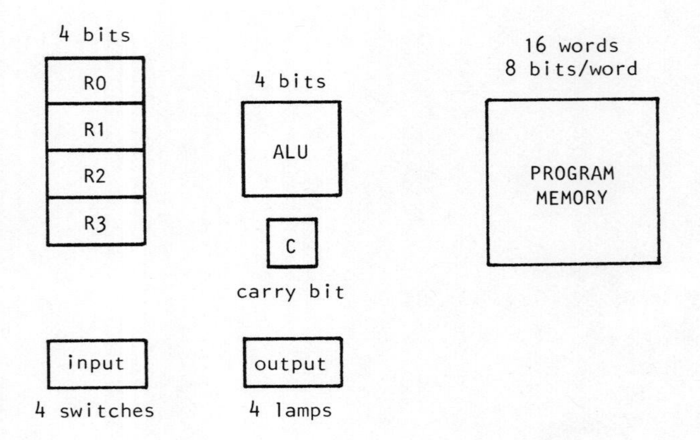

FIGURE D1-1 Processor structure.

shown. C is the contents of the carry flip-flop shown in the block diagram.

TABLE D1-1 Basic Register Instructions

OP	Function
00	$(S) + C_I \rightarrow (D)$
01	$(S) + (D) + C_I \rightarrow (D)$
10	$(D) - (S) - \overline{C_I} \rightarrow (D)$

TABLE D1-2 Carry Bit

CC	C_I
00	0
01	1
10	C
11	$\overline{C}$

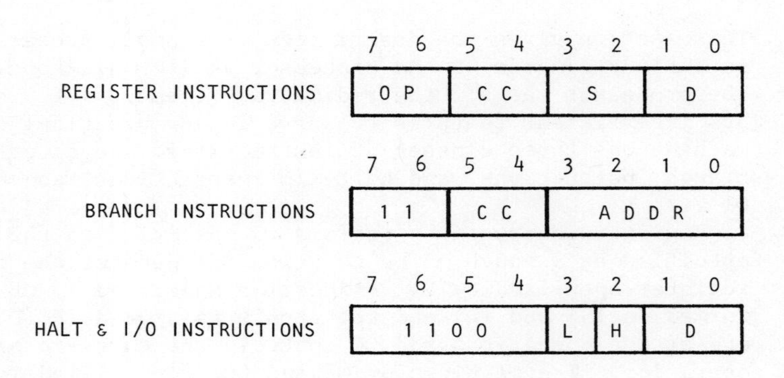

FIGURE D1-2 Instruction formats.

Register instructions also cause the C flip-flop to be loaded. C equals 1 whenever an addition overflows or when a subtraction does not produce a borrow. No other commands affect the C flip-flop. Table D1-3 gives some examples of register instructions.

TABLE D1-3 Extended Register Instructions

MNEMONIC		OPERATION	CODE			
CLEAR	R2	R2 − R2 → R2	1 0	0 1	1 0	1 0
INCREMENT	R3	R3 + 1 → R3	0 0	0 1	1 1	1 1
ROTATE	R1	R1 + R1 → R1	0 1	0 0	0 1	0 1
ADD	R1, R0	R1 + R0 → R0	0 1	0 0	0 1	0 0
SUBTRACT	R2, R0	R0 − R2 → R0	1 0	0 1	1 0	0 0
MOVE	R3, R2	R3 → R2	0 0	0 0	1 1	1 0

Referring to Figure D1-2, branch instructions cause a branch to ADDR in instruction memory if C_I is equal to 1. Again C_I is defined by CC, unless CC = 0, which indicates another instruction class. This gives us three branch commands: unconditional, conditional on C, and conditional on $\overline{C}$.

Halt and input/output instructions are obtained when OP = 11 and CC = 00, as shown in Figure D1-2. If the L-bit is set, then the instruction loads (inputs) the data from the switch register into the register specified by D. The H-bit specifies that the machine is to halt after execution of the instruction. When the machine halts, the register specified by D is displayed in the output lights. Note that both L and H may equal 1 which causes the register to be loaded and then displayed in the lights.

SAMPLE PROGRAM

The program in Table D1-4 multiplies two 4-bit unsigned numbers to produce an 8-bit product. Note that the multiplicand is set in the switch register before the program is started. When the program halts the first time, the multiplier is set in the switches and then the machine is restarted. When it halts again the lights display the most significant four bits of the product, and when restarted it will halt displaying the other four bits. You should check over the program to verify your understanding of the instruction set.

TABLE D1-4 Sample Program

ADDRESS (BASE-4)	CONTENTS (BASE-4)	OPERATION	COMMENTS
0 0	3 0 3 1	SW → R1, HALT	X → R1
0 1	2 1 0 0	R0 - R0 → R0	Clear R0
0 2	2 0 1 0	R0 - R1 - 1 → R0	-(X + 1) → R0
0 3	3 0 2 1	SW → R1	Y → R1
1 0	2 1 2 2	R2 - R2 → R2	Clear product
1 1	2 1 3 3	R3 - R3 → R3	
1 2	0 1 0 0	R0 + 1 → R0	Increment R0 and test for zero
1 3	3 2 2 3	BR 23 if C = 1	
2 0	1 0 1 2	R1 + R2 → R2	Add Y to product
2 1	0 2 3 3	R3 + C → R3	
2 2	3 1 1 2	BR 12	Loop back
2 3	3 0 1 3	R3 → LIGHTS, HALT	Display product
3 0	3 0 1 2	R2 → LIGHTS, HALT	
3 1	3 1 0 0	BR 0 0	Start over
3 2			(unused)
3 3			

ASSIGNMENT

Implement the simple processor described above. A block diagram for the implementation is suggested in Figure D1-3.

A three phase system clock is suggested. Register loading and counting occurs at the end of each clock phase. Different operations occur in each phase as described below.

Register instructions:

Phase 1 - Load (S) into RS.

Phase 2 - Load (D) into RD.

Phase 3 - Compute function of RS and RD, store result into (D), and increment PC.

Branch instructions:

Phase 1, 2 - Unused.

Phase 3 - Load PC if C_I = 1, otherwise increment PC.

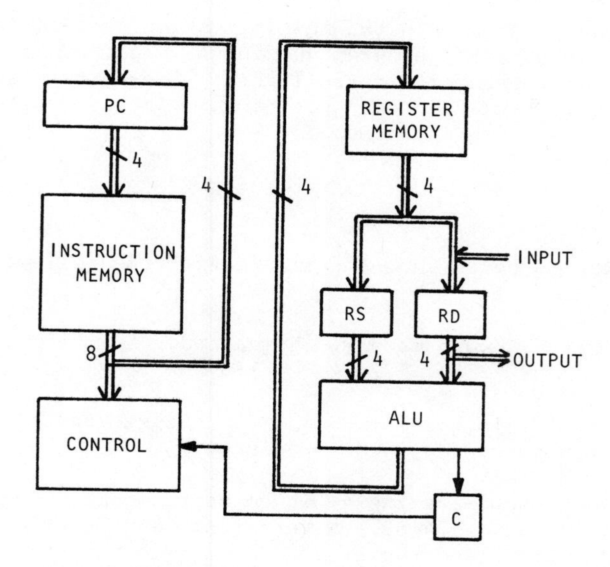

FIGURE D1-3 Block diagram.

Halt and I/O instructions:

Phase 1 - Unused.

Phase 2 - If L = 1, load the switch register into RD.

Phase 3 - If L = 1, store RD into (D). Increment PC.
If H = 1, halt.

Your system should have the following inputs and outputs:

RESET - A push button to initialize control and clear PC.

START - A push button that starts program execution at PC.

INCPC - A push button to increment PC when the machine is halted.

INSTR - Eight toggle switches used to specify an instruction.

LOAD - A push button that loads the instruction memory word addressed by PC with INSTR.

SW - The switch register, four toggle switches used to input data to a program. SW may share the same physical switches with INSTR.

DATA OUT - Four lamps that display the contents of RD at all times.

PC OUT - Four lamps that display the contents of PC at all times.

Test your system with the multiplication program in Table D1-4. Then write and test a program that inputs a pattern from the switch register and rotates it in the lights. The pattern should spend an equal amount of time in each position. Finally, write and test a program of your own choosing.

PARTS

In addition to parts in the standard kit, the following ICs will be needed:

1 - 4-bit × 4-word memory. Suggested: 1 × 74170. Alternates: 1 × 7489, 2 × 74172, 4 × 74173.

2 - 4-bit registers for RS and RD. Suggested: 2 × 74175. Alternates: any 4-bit registers.

1 - Synchronous-loading 4-bit counter for PC. Suggested: 1 × 74161 or 74163. Alternates: any 4-bit counter and edge-triggered register.

1 - 8-bit by 16-word instruction memory. Suggested: 2 × 7489.

1 - 74157 2-input, 4-bit multiplexer in addition to the one in the standard kit.

HINTS

The ALU functions of Table D1-2 can be implemented with a 7483 4-bit adder and appropriate circuitry to select true, complement, or zero inputs. Take advantage of the clear and complement capabilities of the 74175 registers or use a 7487 4-bit true/complement/one/zero element.

The temporary registers RS and RD are required because it is not possible to simultaneously read two words and write one word in the 74170. If edge-triggered registers such as 74173s are used, one temporary register can be eliminated, and with 74172s both temporary registers can be eliminated. The corresponding extra clock phases can also be eliminated, so that with 74172s a single phase clock can be used. The 74172 also eliminates the need for an extra 74157 multiplexer.

Program counter loading for branches must be synchronous. Why?

D2 - ASSOCIATIVE MEMORIES

SUMMARY

This project is an application of associative memories. A small memory system is designed that can be quickly searched for maximum or minimum entries. The search time depends on the width of the memory words rather than on the number of words in the memory.

INTRODUCTION

This project uses associative memories such as the Intel 3104. As shown in Figure D2-1, the 3104 is organized as 4 words by 4 bits. The functions of the 3104s input and output lines are summarized below.

$\overline{A}_0\overline{A}_1\overline{A}_2\overline{A}_3$ - Each of these active-low input lines selects one of the 4-bit words. More than one word may be selected.

$\overline{D}_0\overline{D}_1\overline{D}_2\overline{D}_3$ - The complements of these four bits of input data are stored during write operations and are used for comparison during match operations. The storage cells are D-latches.

$\overline{E}_0\overline{E}_1\overline{E}_2\overline{E}_3$ - Each of these active-low input lines enables a column of bits for writing or matching operations.

$\overline{WE}$ - When this input lines is low, the enabled bits of the selected words are loaded. All other bits are unchanged. Bit j of word i (i.e., m_{ij}) is loaded with $\overline{D}_j$ when $\overline{A}_i = 0$, $\overline{E}_j = 0$, and $\overline{WE} = 0$.

$\overline{O}_0\overline{O}_1\overline{O}_2\overline{O}_3$ - The complement of the stored data in the selected words is always present on these output lines. That is,

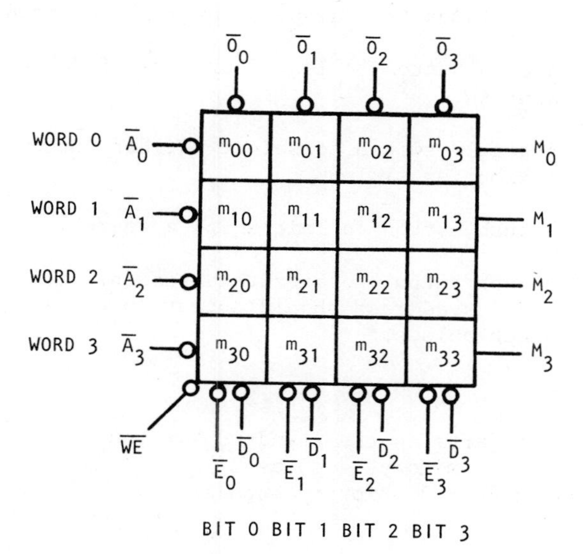

FIGURE D2-1 Intel 3104 16-bit associative memory.

$$\overline{O}_j = \overline{\overline{A}_0 m_{0j} + \overline{A}_1 m_{1j} + \overline{A}_2 m_{2j} + \overline{A}_3 m_{3j}}$$

$M_0M_1M_2M_3$ - Each of these four active-high output lines indicates a match of the selected bits of the corresponding stored word and the complement of the input words. That is,

$$M_i = (\overline{E}_0 + (m_{i0} \oplus \overline{D}_0))(\overline{E}_1 + (m_{i1} \oplus \overline{D}_1))(\overline{E}_2 + (m_{i2} \oplus \overline{D}_2))(\overline{E}_3 + (m_{i3} \oplus \overline{D}_3))$$

All outputs are the open-collector type and require a pull-up resistor.

ASSIGNMENT

You are to design and construct an 8-word by 4-bit memory system that can be quickly searched for maximum or minimum entries. The search should use an algorithm whose execution time depends on the number of bits in each word and not on the number of words to be searched.

Your circuit should have the following inputs and outputs:

ADDR IN - Three toggle switches to select one of the eight words to load or display.

ADDR OUT - Three lamps that display the selected address.

DATA IN - Four toggle switches for loading data into the memory.

DATA OUT - Four lamps that display the selected word at all times.

LOAD - A push button to load four bits of data into the selected address.

MAX/MIN - A toggle switch to select whether the search is for the maximum or minimum.

INIT - A push button to initialize a search.

CLOCK - A push button to execute a search. The maximum or minimum and its address should appear in the lamps after four clock pulses.

PARTS

In addition to parts in the standard kit you will need an 8-word by 4-bit associative memory, such as 2 × Intel 3104. A 3-to-8 decoder (7442), and an 8-to-3 priority encoder (74148) may also be useful.

D3 - SCOPE-A-SKETCH

SUMMARY

A system that can selectively store and erase any pattern of dots in a 64 × 64 matrix is designed in this project. The stored pattern is displayed on an oscilloscope.

SYSTEM DESCRIPTION

The object of this project is to design and build a system that can create and display any pattern of dots in a 64 × 64 matrix. A block diagram of the system is given in Figure D3-1. The heart of the system is a 4096-bit shift register and a 12-bit counter. At each clock pulse, the shift register is shifted one position and the counter is incremented. The lower six bits of the counter are interpreted as an X position and the high six bits are interpreted as a Y position. These bits are connected to two digital-to-analog converters (DACs); the resulting X and Y voltages are connected to a display oscilloscope. Hence, the counting action of the counter causes a beam to scan horizontally across the display, moving vertically one line after each scan. At each count, a new bit appears at the output of the shift register. This bit is fed to the Z-axis of the scope to

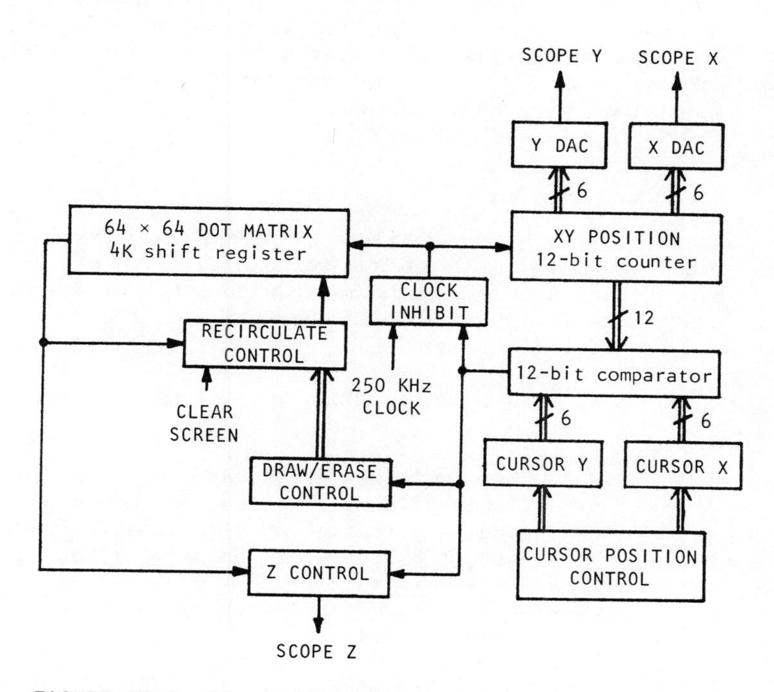

FIGURE D3-1 Block diagram.

determine whether or not the corresponding position of the display is to be intensified (i.e., have a dot appear) or to be left blank. The clock rate is chosen so that a complete shift register and counter cycle is made about 60 times per second, so that a flicker-free display is seen by the user.

A few features are needed to make the system practical. The user needs a way of clearing the entire display and of selectively drawing or erasing dots at any position on the screen. Clearing the screen can easily be accomplished by a "clear screen" control that causes the shift register to be loaded with all zeroes (after 4096 shifts). To draw or erase a particular dot the user must first specify which dot is to be modified. For this purpose a cursor is provided. The X and Y positions of the cursor are set by the user via the "cursor position control." When the user requests a "draw," the current scope XY position is compared with the cursor position. When the positions become equal, the corresponding dot position is available at the shift register and the "draw/erase control" and "recirculate control" cause a 1 to be written instead of the old contents. Erases are similar, with a 0 being written.

The user must have some means of knowing where the cursor is positioned. The easiest scheme to implement is to require the user to enter the cursor position as a 12-bit number and let the user interpret the number as a position. However, this is not too convenient for the user. It is better to provide some means for the user to see the cursor directly on the display. The suggested way of accomplishing this is shown in Figure D3-1. Two counters are used to store the cursor X and Y coordinates. The user, through the "cursor position control," can change either of these counters to move the cursor. The current cursor position is displayed by inhibiting the clock for four cycles when the scope beam is at the cursor position. The corresponding dot is intensified for four cycles, making it brighter than other dots on the screen.

ASSIGNMENT

You are to design and build a scope-a-sketch system with the capabilities outlined here. You are welcome to deviate from the suggested block diagram if you can come up with a better system. The suggested system has three outputs, to the X, Y, and Z axes of the display scope, and the following controls:

CLEAR - This switch, when set, clears the entire screen.

UP, DOWN, LEFT, RIGHT - These switches are used to move the cursor. As long as a switch is set, the cursor should move in the corresponding direction at a reasonable rate, say 8 to 16 dots (1/8 to 1/4 of the screen) per second.

DRAW - When this switch is set, a dot is drawn at the current cursor position.

ERASE - When this switch is set, a dot is erased at the current cursor position. When both DRAW and ERASE are set, the dot should be complemented.

Note that a DRAW or ERASE command can be given while the cursor is moving, and so the matrix must be updated on the fly. What does this imply about the relationship between the maximum cursor speed and the refresh rate?

PARTS

In addition to parts in the standard kit, the following ICs will be needed:

1 - 4K shift register. Recommended: 2 × Intel 2401 or 4 × Intel 2405.

1 - 12-bit position counter for scan. Recommended: 3 × any 4-bit counter.

2 - 6-bit up/down counters for cursor. Recommended: 4 × 74193 or 74191.

1 - 12-bit comparator. Recommended: 3 × 7485. Alternate: 2 × LM311 or other analog comparator.

HINTS

The Z-axis of most oscilloscopes is DC coupled, but some are AC coupled. With an AC-coupled Z-axis, problems may be encountered when trying to display a completely blank dot matrix, and you will have to design circuitry to overcome this.

One of the most critical parts of the design is the clock inhibit for the cursor display. Be careful not to introduce any hazards here.

OPTIONS

There are other ways to implement the cursor display, for example an intensified caret can be displayed instead of a dot. A blinking cursor is very easy to display (perhaps too easy since it makes the project complexity closer to C-series).

The cursor position control can be implemented with two potentiometers for X and Y instead of two counters and push buttons. The potentiometers produce analog output voltages that can be compared with the scanning position using analog comparators. This approach requires less circuitry than the completely digital approach.

The size of the system can be reduced by displaying a smaller dot matrix. For example, a 32 × 32 matrix requires only 1K of memory and 10 bits of position information.

It is not necessary to use shift registers to store the dot matrix. Static random-access memories such as 2102s (1K by 1 bit) can be used instead, by connecting the output of the position counter to the address inputs of the memory. A dynamic 4K memory such as the 2107 can also be used if its refresh requirements are met.

D4 - FLOATING-POINT MULTIPLIER

SUMMARY

In this project a simple floating-point multiplier will be built that accepts as operands two 12-bit, possibly unnormalized floating-point numbers, and that outputs a 12-bit normalized number and an overflow indicator.

INTRODUCTION

The floating-point format used in this project is as follows.

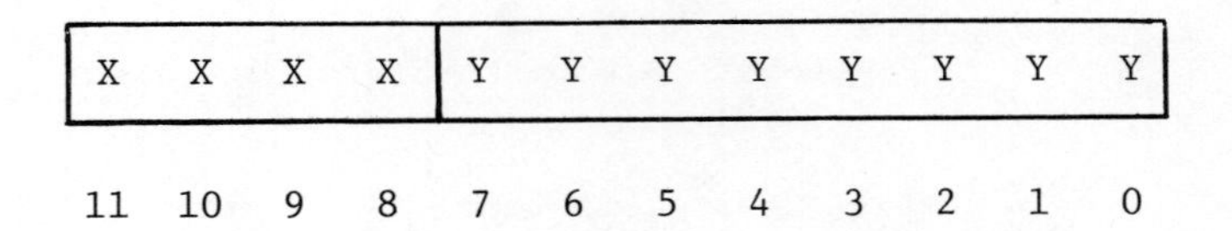

The *mantissa* of the number is an 8-bit unsigned positive number, M = .YYYYYYYY, with the radix point on the left. The *exponent* of the represented number is a 4-bit unsigned positive number, E = XXXX. The value of the represented number is $M*2^E$. A number in this format is *normalized* if the most significant bit of the mantissa is 1. Some examples of normalized numbers in this format are given below.

1111 11111111 = $(255/256)*2^{15}$ = $255*2^7$ (largest number)

0000 10000000 = $(128/256)*2^0$ = ½ (smallest normalized number)

0001 10000000 = $(128/256)*2^1$ = 1

0101 10001000 = $(136/256)*2^5$ = 17

Some examples of unnormalized numbers are as follows.

0010 01000000 = $(64/256)*2^2$ = 1

1000 00010001 = $(17/256)*2^8$ = 17

0000 00000000 = $(0/256)*2^0$ = 0 ("clean" zero)

0101 00000000 = $(0/256)*2^5$ = 0 (a "dirty" zero)

0001 00001011 = $(11/256)*2^1$ = 22/256

Note that there are many possible representations of zero, but only one of them is defined to be "clean." An attempt to normalize the smallest representable nonzero numbers (in the range $½ > N > 0$) results in *exponent underflow*.

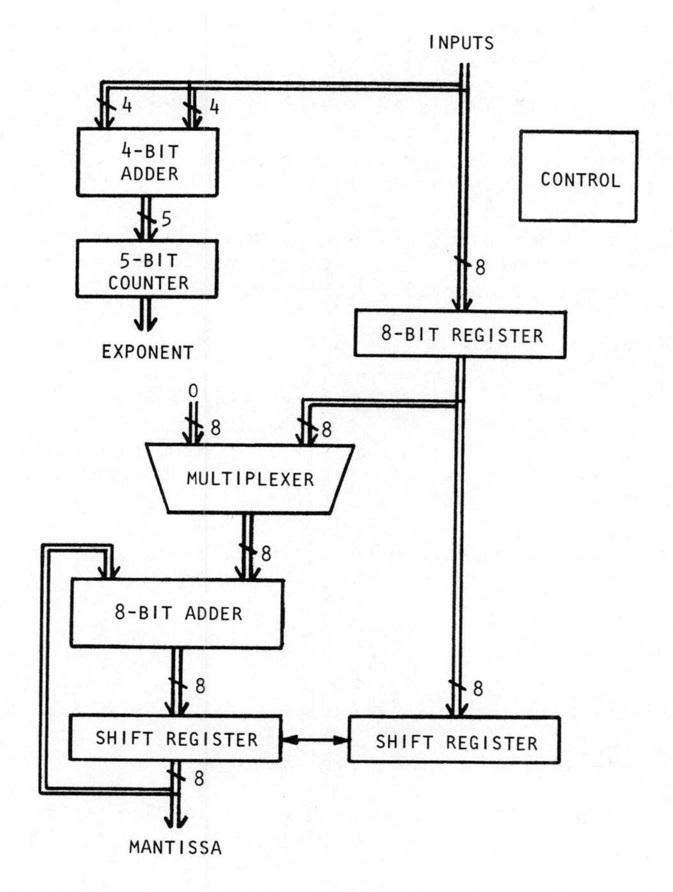

FIGURE D4-1 Block diagram.

ASSIGNMENT

Figure D4-1 gives a suggested block diagram for a multiplier for the floating-point numbers described above. You are to design the multiplier circuit subject to the following requirements.

1) There should be two push buttons, PB1 and PB2. PB1 is used to reset the system and initialize, and PB2 loads segments of operands as follows:

    a) On the first depression, the 8-bit mantissa of the first operand is loaded from eight toggle switches.

    b) On the second depression, the 8-bit mantissa of the second operand is loaded.

    c) On the third depression, the two 4-bit exponents are loaded. Execution proceeds automatically and the result is displayed in 12 lamps.

2) Use a conventional add-and-shift algorithm for the multiplication of the mantissas.

3) Overflow should be indicated by a lamp.

4) All true zero results are to be converted into "clean zero" form.

5) Input operands may be unnormalized, but the result must always be normalized if it is nonzero. However, you may assume that exponent underflow will never occur.

Answer the following questions:

1) Why was a 5-bit counter specified in the block diagram to hold the exponent sum?

2) Can underflow occur if the input operands are constrained to be normalized? Explain.

3) What techniques could be used to make the multiplier operate faster?

PARTS

In addition to parts in the standard kit, you will need the following.

1 - 8-bit 2-input multiplexer. Suggested: 2 × 74157.

1 - 8-bit adder. Suggested: 2 × 7483 or 74283. Alternate: Replace adder and multiplexer with 2 × 74181, 74281, or 74381.

2 - 8-bit parallel-in bidirectional shift register. Suggested: 4 × 74194.

1 - 8-bit register. Suggested: 2 × 74174 or 74175.

1 - 5-bit parallel load counter. Suggested: 2 × 74161, 74191, or 74193.

HINTS

The control circuitry is the most difficult part of this project, since the data paths are given. Do a clean design of a sequential machine to cycle through the control states needed to automatically execute your floating-point multiplication algorithm.

OPTIONS

Use a more complex floating-point representation, such as the one given in Project E3.

If you wish, you may handle exponent underflow. Results that produce exponent underflow should be automatically converted to "clean" zero, and an indicator should be lit.

REFERENCES

Floating-point representations and hardware are discussed in Gschwind and McCluskey [1975], Stone [1972], Booth [1971], and most other texts on computer organization.

D5 - THE GAME OF DIB

SUMMARY

DIB is a game for two players. A machine can be designed to play DIB with a human opponent. The machine uses an optimal strategy, so that the machine will always win if its opponent makes any mistakes. The machine's strategy can be implemented with a small sequential circuit. The objective of this project is to design the strategy circuit and in addition a nice user interface that allows a human to play DIB with the machine.

RULES OF DIB

The game of DIB starts with a single heap containing an odd number of sticks. The two players alternately remove sticks from the heap and collect them. At each turn a player must remove 1, 2, 3, or 4 sticks. The object of the game is to have collected an even number of sticks when all sticks are removed from the heap. Since the original number of sticks in the heap is odd, one of the players must win.

ANALYSIS

A player is in a winning position if, regardless of what his opponent does, he can always answer with a move bringing him into a new winning position. The following positions are winning positions for DIB (assume $k \geq 0$).

A_k have an even number after the move, and $0 + 6k$ left in the heap.

B_k have an even number after the move, and $1 + 6k$ left in the heap.

C_k have an odd number after the move, and $5 + 6k$ left in the heap.

That is, you have a winning position if you have an even number after your move, and there are 0 or 1 modulo 6 sticks left in the heap, or if you have an odd number, and there are 5 modulo 6 sticks left.

The fact that these are winning positions can be proved by induction on k. Proof of the base step (k=0) is as follows.

A_0 This is the end winning position.

B_0 The opponent must take the last stick, putting him in the end losing position.

C_0 Let the opponent remove i sticks, so that there are 5-i sticks left. If i is even, take all the remaining sticks, leaving you in position A_0. If i is odd, take all but 1, leaving you in position B_0.

In the induction step we assume that the proposition is true for k, and prove that it then must be true for k+1. Proof of the induction step is left as an exercise for the interested reader.

STRATEGY

A good strategy for the game of DIB is as follows.

> Try to get an even number with 0 or 1 modulo 6 sticks left in the heap, or an odd number with 5 modulo 6 sticks left. As soon as such a position is obtained, by proper moves we can keep a winning position until no sticks remain in the heap.

This also shows that if the original number of sticks in the heap is 1 modulo 6 then the starting player has a losing position, that is, he will always lose against an experienced player (or a machine). All other uneven starting values give the first player a chance to win.

The strategy can be described by a state table, as shown in Table D5-1. The states are named 0u,...,5u,0e,...,5e, describing the number of sticks modulo 6 left in the heap (0-5) and whether the machine has an even or uneven number after its move (e or u). For each state and number of sticks that the user may take, the table shows how many sticks the machine must take if it can obtain a winning position. Notice that for the winning positions (5u, 0e, 1e) the machine can always obtain a new winning position regardless of the user move. By taking moves corresponding to the blanks in the table, the user can avoid being beaten by the machine. The way that the blanks are filled in determines how the machine defends itself when it is in a losing position. The blanks can also be filled in to minimize the number of states of the reduced state table (the minimum is 5 states: 5u, 0e, 1e, 3e, and 5e).

ASSIGNMENT

You are to design and build a machine that plays DIB against a human opponent. The machine should use the strategy described above. It

TABLE D5-1 State Table for DIB

	User Takes			
State	1	2	3	4
0u		1e,3	0e,3	1e,1
1u			1e,3	0e,3
2u	0e,1			1e,3
3u	1e,1	0e,1		
4u	0e,3	1e,1	0e,1	
5u[a]	1e,3	0e,3	1e,1	0e,1
0e[a]	1e,4	0e,4	1e,2	0e,2
1e[a,b]	5u,1	1e,4	0e,4	1e,2
2e		5u,1	1e,4	0e,4
3e[b]	0e,2		5u,1	1e,4
4e	1e,2	0e,2		5u,1
5e[b]	0e,4	1e,2	0e,2	
	Next State, Machine Takes			

[a]These are the winning positions for the machine.

[b]Starting states, depending on the number of sticks in the original heap.

should be possible to initialize the machine with any odd number of sticks from 1 to 63. The machine keeps track of the total number of sticks left in the heap and whether it has collected an odd or even number of sticks. After each user move the machine picks and executes its own best move. When the number of sticks in the heap becomes zero, the machine indicates the winner.

Your machine should have the following inputs and outputs:

LOAD - This pushbutton loads the starting number of sticks from the toggle switches and initializes the state of the machine.

STICKS - These five toggle swithces indicate the starting number of sticks, an odd number between 1 and 63.

USER - This push button is used to indicate the user's move. It is pushed once for each stick the user wishes to take. This switch is ignored after four pushes or if there are no sticks left in the heap.

MACHINE - This push button is used to tell the machine that the user has finished taking sticks and is ready for the machine to move. This button should be ignored unless USER has been pushed at least once.

STICKSLEFT - These six lamps display in binary the number of sticks left in the heap at all times. This output is updated every time USER is pushed, and after MACHINE is pushed.

MACHINE MOVE - These three lamps indicate in binary the number of sticks that the machine takes. They are updated every time MACHINE is pushed.

WINNER - These two lamps, both off during the game, are activated at the end of the game to indicate the winner (one lamp for machine, one for user). Depending on the winner, the winner's lamp is turned on in response to a push of either USER or MACHINE.

PARTS

In addition to the standard kit you will need the following:

1 - 6-bit binary down-counter. Recommended: 2 × 74191 or 74193. Alternate: 2 × any 4-bit binary counter.

HINTS

Rather than calculate from the 6-bit counter the number of sticks left modulo 6 for each machine move, maintain a separate modulo 6 counter that gives this number at all times. You will still be required to properly initialize the modulo 6 counter when LOAD is pushed.

OPTIONS

Many variations of the input/output specifications are possible. The important thing is to provide a nice interface to the human user. It is not sufficient to simply build the reduced 5-state strategy machine from Table D5-1.

You may wish to provide a secret output lamp that is lit whenever the machine has a winning next state, corresponding to the filled-in positions of Table D5-1. If this lamp is updated every time USER is pushed, then the user can play a perfect game by always pushing USER just until the lamp goes out. Use this feature to amaze your friends at your skill at DIB.

There is a whole class of games similar to DIB that allow from 1 to n sticks to be taken on each turn (n is fixed for each game). For even n, the winning positions are:

A_k have an even number after the move, and $0 + (n+2)k$ left in the heap.

B_k have an even number after the move, and $1 + (n+2)k$ left in the heap.

C_k have an odd number after the move, and $1 + n + (n+2)k$ left in the heap.

For odd n, the winning positions are:

A_k have an even number after the move, and $0 + 2(n+1)k$ left in the heap.

B_k have an even number after the move, and $1 + 2(n+1)k$ left in the heap.

C_k have an odd number after the move, and $n+1 + 2(n+1)k$ left in the heap.

D_k have an odd number after the move, and $n+2 + 2(n+1)k$ left in the heap.

Even the n=2 game is interesting; you may wish to build a machine for one of these other games.

ACKNOWLEDGEMENT

The strategies for the class of DIB-like games were derived by Professor Dag Belsnes.

D6 - PROGRAMMABLE MUSIC SYNTHESIZER

SUMMARY

In this project you will design and build a system that can play any sequence of 512 notes spanning four musical octaves. The system has a 512-word memory to store binary encodings of the musical notes and circuits to generate musical notes corresponding to the encodings. A 512-word score for "The Flight of the Bumble Bee" is given as an example.

INTRODUCTION

Table D6-1 lists the 12 musical steps that form an octave. Each step has a frequency that exceeds the frequency of the note below it by the twelfth root of two, or approximately 1.059. Each octave is separated by a factor of two. Middle C is located at 261 Hz. Hence the C below middle C is at 130.5 Hz, C above middle C is at 522 Hz, and high C is at 1044 Hz. If we have a signal with a frequency corresponding to a note in one octave, we can obtain the same note in lower octaves by dividing the frequency of the given signal by a power of two.

TABLE D6-1 Music Codes

CODE		NOTE	FREQ (Hz)	PERIOD (μsec)	PROM CODE		PROM CODE (hex)
0	0000	SILENCE	--		1111	1111	FF
1	0001	C	130.5	7662	0001	0000	10
2	0010	C#	138.5	7232	0001	1101	1D
3	0011	D	146.5	6826	0010	1010	2A
4	0100	D#	155	6444	0011	0110	36
5	0101	E	164.5	6082	0100	0001	41
6	0110	F	174	5740	0100	1100	4C
7	0111	F#	184.5	5418	0101	0110	56
8	1000	G	195.5	5114	0101	1111	5F
9	1001	G#	207	4828	0110	1000	68
10	1010	A	219.5	4556	0111	0001	71
11	1011	A#	232.5	4300	0111	1001	79
12	1100	B	246.5	4060	1000	0000	80
13	1101	TEST	125	8000	0000	0101	05
14	1110	TEST	125	8000	0000	0101	05
15	1111	SILENCE	--	--	1111	1111	FF

The programmable music synthesizer allows the user to specify a sequence of 512 notes to be played. Each note is given by a six-bit code; four bits specify one of the 12 basic notes or silence, and two bits specify one of four octaves. Hence the musical program is stored in a 512-word by 6-bit memory.

In order to explain how musical tones are generated by the synthesizer, it is easiest to start with the output tone and work backwards (Refer to Figure D6-1). The output signals produced by digital circuits are square waves and audio square waves are not too pleasing to listen to. Something closer to a sine wave is desirable. The circuit shown in Figure D6-2 can be used to convert an input square wave of frequency $3f$ into an output wave at frequency f that approximates a sine wave.* In this circuit the exclusive OR gate and two flip-flops form a modulo-3 counter with a symmetric output; if the input frequency is $3f$, then the output of the second flip-flop is a square wave with frequency f. Two resistors form a weighted summing network that adds the $3f$ signal

*John Taylor, "Frequency divider plus op amp approximates sine wave," *Electronics*, vol. 48, no. 2 (January 23, 1975), p. 89.

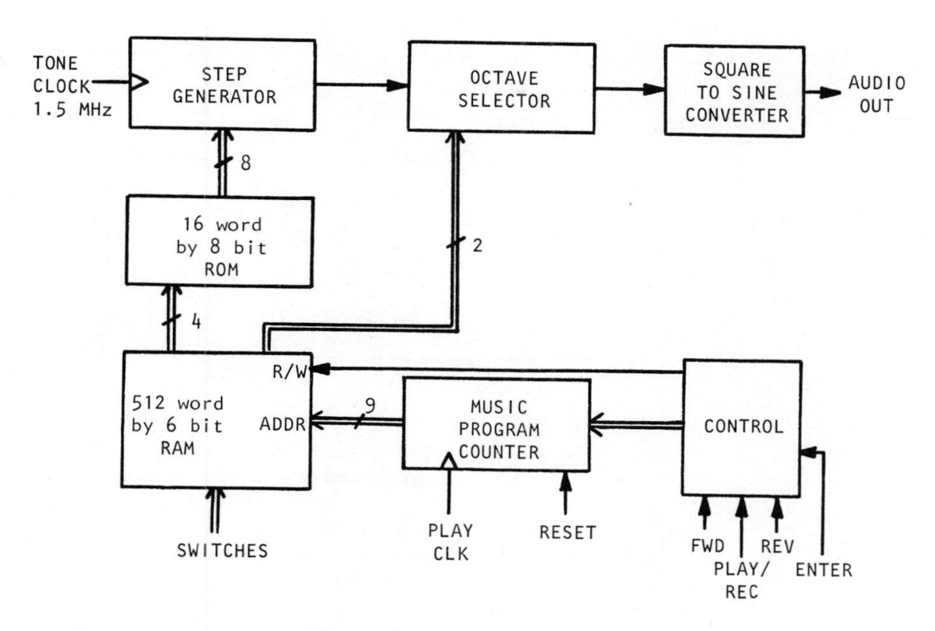

FIGURE D6-1 Block diagram.

and the f signal. The $3f$ signal is out of phase with the f signal and cancels its third harmonic. Only the fifth and higher odd harmonics remain to be filtered by a capacitor. The resulting signal, despite appearances, sounds reasonably sinusoidal to the ear, with the remaining harmonics giving the sound some character.

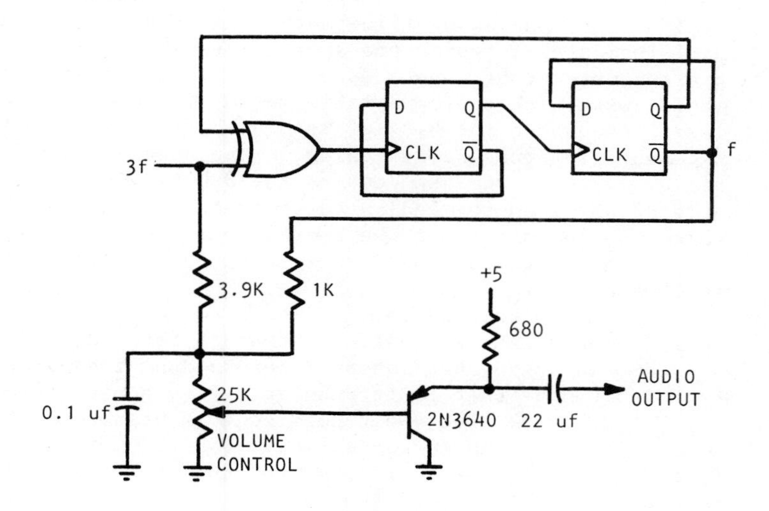

FIGURE D6-2 Square-to-sine converter.

TABLE D6-2 Octave Codes

OCTAVE CODE	OCTAVE SYMBOL	OCTAVE	COMMENTS
00	=	130.5-246.5 Hz	
01	-	261-493 Hz	MIDDLE C = 261 Hz
10		522-986 Hz	
11	+	1044-1972 Hz	1 KHz TEST SIGNAL IN THIS OCTAVE

We see above that we need to input a signal of frequency $3f$ to the square-to-sine converter to get an output frequency of f. Octave selection can be accomplished by a binary counter and a multiplexer. If we apply a frequency of $48f$ to the clock input of a 4-bit binary counter, the four output bits of the counter will be square waves with frequencies of $24f$, $12f$, $6f$, and $3f$. Each counter output gives a signal in a different octave. For example, if the input frequency to the counter is 6264 Hz, the output frequencies are 3132 Hz, 1566 Hz, 783 Hz, and 391.5 Hz, and square-to-sine processing results in frequencies of 1044 Hz, 522 Hz, 261 Hz, and 130.5 Hz, corresponding to the note C in each of the four octaves.

Referring to Figure D6-1, we see then that for each note the "step generator" should produce a signal with frequency 48 times the frequency of that note in the lowest octave. This frequency generation can be done by simple counters, since the highest frequency to be generated, about 12 KHz, is far below what can be handled by standard digital circuits. A "period counter" is loaded with a number whose complement is proportional to the period of the desired signal. The counter counts up until it overflows, then it is reloaded and counts again. The counter overflow output has the desired frequency. Table D6-1 indicates for each the appropriate value to be loaded into an 8-bit counter that counts up to 11111111 and then overflows. (Experimental work with a programmable music synthesizer has shown that at least seven or eight bits of resolution for the period counter are necessary for reasonable tuning of the notes.) With a reference signal of 1.5 MHz (.66 μsec/count) the 8-bit counter has an overflow rate of 48 times the indicated frequency. The 8-bit counter initial values for each note may be stored in a 16-word by 8-bit ROM, so that corresponding to each 4-bit note code there is a corresponding value to look up in the ROM.

In summary, the sequency of notes to be played is stored in a 512-word by 6-bit RAM; sequential notes are fetched from the RAM and played at a speed that can be determined by a separate "play clock." Each 6-bit code that is fetched specifies a note (4 bits) and an octave (2 bits). The 4-bit note code is used to select from a ROM an 8-bit initial value for a counter. The counter produces an output frequency 48 times the frequency of the selected note in the lowest octave. A 4-bit counter then divides to obtain 24, 12, 6, and 3 times this note's frequency, and one of these signals is selected according to the 2-bit octave code. This signal is finally divided by three and converted to a pseudo-sinusoidal audio signal by a square-to-sine converter.

ASSIGNMENT

You are to build a programmable music synthesizer following the block diagram of Figure D6-1. In addition to the playback circuitry described already, your system obviously must also have circuits that allow a music program to be entered. The system should have the following inputs and outputs.

PLAY/REC - A toggle switch to determine the mode of operation, playing or recording a program. In play mode the program should run continuously, looping at the end of the memory.

FWD - A push button that increments the music program counter in REC mode.

REV - A push button that decrements the music program counter in REC mode.

ENTER - A push button for loading the current music program location with a 6-bit code in REC mode.

CODE IN - Six toggle switches used above.

CODE OUT - Six lamps that display the music program memory output at all times.

RESET - A push button that resets the music program counter to location zero.

MEMFULL - A lamp that indicates that the music program counter has reached its highest value.

VIBES - The audio output.

You will need a 1.5 MHz signal source for the tone clock. This source should preferably be variable, for tuning the synthesizer. You will also need a variable source for the play clock, with a range of at least 2 Hz to 15 Hz. Tune your system (the tone clock) using a calibrated signal generator or someone with perfect pitch. The ROM values in Table D6-1 were adjusted by the latter method. Peculiarities in a particular system design may require further adjustment of the ROM values, for reasons such as overhead in reloading the step counter, always counting one time too few or too many, and so on. Test the controls of your system and then give it (and your fingers) a real test by entering the sample music program shown in Table D6-3. It took the original designer about a half hour to enter by hand.

TABLE D6-3 "The Flight of the Bumble Bee" by N. Rimsky Korsakoff, arranged by Bruce P. Almich

ADDR*	NOTE	CODE	ADDR	NOTE	CODE	ADDR	NOTE	CODE	ADDR	NOTE	CODE
000	E +	0101 11	060	E -	0101 01	140	E -	0101 01	220	A -	1010 01
001	D# +	0100 11	061	D# -	0100 01	141	D# -	0100 01	221	G# -	1001 01
002	D +	0011 11	062	D -	0011 01	142	D -	0011 01	222	G -	1000 01
003	C# +	0010 11	063	C# -	0010 01	143	C# -	0010 01	223	F# -	0111 01
004	D +	0011 11	064	C -	0001 01	144	D -	0011 01	224	G -	1000 01
005	C# +	0010 11	065	F -	0110 01	145	C# -	0010 01	225	F# -	0111 01
006	C +	0001 11	066	E -	0101 01	146	C -	0001 01	226	F -	0110 01
007	B	1100 10	067	D# -	0100 01	147	B =	1100 00	227	E -	0101 01
010	C +	0001 11	070	E -	0101 01	150	C -	0001 01	230	F -	0110 01
011	B	1100 10	071	D# -	0100 01	151	C# -	0010 01	231	F# -	0111 01
012	A#	1011 10	072	D -	0011 01	152	D -	0011 01	232	G -	1000 01
013	A	1010 10	073	C# -	0010 01	153	D# -	0100 01	233	G# -	1001 01
014	G#	1001 10	074	C -	0001 01	154	E -	0101 01	234	A -	1010 01
015	G	1000 10	075	C# -	0010 01	155	F# -	0111 01	235	A# -	1011 01
016	F#	0111 10	076	D -	0011 01	156	G -	1000 01	236	A -	1010 01
017	F	0110 10	077	D# -	0100 01	157	G# -	1001 01	237	G# -	1001 01
020	E	0101 10	100	E -	0101 01	160	A -	1010 01	240	A -	1010 01
021	D#	0100 10	101	D# -	0100 01	161	G# -	1001 01	241	G# -	1001 01
022	D	0011 10	102	D -	0011 01	162	G -	1000 01	242	G -	1000 01
023	C#	0010 10	103	C# -	0010 01	163	F# -	0111 01	243	F# -	0111 01
024	D	0011 10	104	C -	0001 01	164	F -	0110 01	244	G -	1000 01
025	C#	0010 10	105	F -	0110 01	165	A# -	1011 01	245	F# -	0111 01
026	C	0001 10	106	E -	0101 01	166	A -	1010 01	246	F -	0110 01
027	B -	1100 01	107	D# -	0100 01	167	G# -	1001 01	247	E -	0101 01
030	C	0001 10	110	E -	0101 01	170	A -	1010 01	250	F -	0110 01
031	B -	1100 01	111	D# -	0011 01	171	G# -	1001 01	251	F# -	0111 01
032	A# -	1011 01	112	D -	0011 01	172	G -	1000 01	252	G -	1000 01
033	A -	1010 01	113	C# -	0010 01	173	F# -	0111 01	253	G# -	1001 01
034	G# -	1001 01	114	C -	0001 01	174	F -	0110 01	254	A -	1010 01
035	G -	1000 01	115	C# -	0010 01	175	F# -	0111 01	255	A# -	1011 01
036	F# -	0111 01	116	D -	0011 01	176	G -	1000 01	256	A -	1010 01
037	F -	0110 01	117	D# -	0100 01	177	G# -	1001 01	257	G# -	1001 01
040	E -	0101 01	120	E -	0101 01	200	A -	1010 01	260	A -	1010 01
041	D# -	0100 01	121	D# -	0100 01	201	G# -	1001 01	261	A =	1010 00
042	D -	0011 01	122	D -	0011 01	202	G -	1000 01	262	A -	1010 01
043	C# -	0010 01	123	C# -	0010 01	203	F# -	0111 01	263	A =	1010 00
044	D -	0011 01	124	D -	0011 01	204	F -	0110 01	264	A -	1010 01
045	C# -	0010 01	125	C# -	0010 01	205	A# -	1011 01	265	A =	1010 00
046	C -	0001 01	126	C -	0001 01	206	A -	1010 01	266	A -	1010 01
047	B =	1100 00	127	B =	1100 00	207	G# -	1001 01	267	A =	1010 00
050	E -	0101 01	130	C -	0001 01	210	A -	1010 01	270	A# -	1011 01
051	D# -	0100 01	131	C# -	0010 01	211	G# -	1001 01	271	A# =	1011 00
052	D -	0011 01	132	D -	0011 01	212	G -	1000 01	272	A# -	1011 01
053	C# -	0010 01	133	D# -	0100 01	213	F# -	0111 01	273	A# =	1011 00
054	D -	0011 01	134	E -	0101 01	214	F -	0110 01	274	A# -	1011 01
055	C# -	0010 01	135	F -	0110 01	215	F# -	0111 01	275	A# =	1011 00
056	C -	0001 01	136	E -	0101 01	216	G -	1000 01	276	A# -	1011 01
057	B =	1100 00	137	D# -	0100 01	217	G# -	1001 01	277	A# =	1011 00

*Addresses are in octal.

TABLE D6-3 (continued)

ADDR	NOTE	CODE	ADDR	NOTE	CODE	ADDR	NOTE	CODE	ADDR	NOTE	CODE
300	A -	1010 01	360	A -	1010 01	440	D	0011 10	520	E +	0101 11
301	A =	1010 00	361	A =	1010 00	441	D#	0100 10	521	D# +	0100 11
302	A -	1010 01	362	A -	1010 01	442	E	0101 10	522	D +	0011 11
303	A =	1010 00	363	A =	1010 00	443	F	0110 10	523	C# +	0010 11
304	A -	1010 01	364	D	0011 10	444	F#	0111 10	524	D +	0011 11
305	A =	1010 00	365	D -	0011 01	445	F	0110 10	525	C# +	0010 11
306	A -	1010 01	366	D	0011 10	446	E	0101 10	526	C +	0001 11
307	A =	1010 00	367	D -	0011 01	447	D#	0100 10	527	B	1100 10
310	A# -	1011 01	370	D#	0100 10	450	D	0011 10	530	C +	0001 11
311	A# =	1011 00	371	D# -	0100 01	451	D#	0100 10	531	B	1100 10
312	A# -	1011 01	372	D#	0100 10	452	E	0101 10	532	A#	1011 10
313	A# =	1011 00	373	D# -	0100 01	453	F	0110 10	533	A	1010 10
314	A# -	1011 01	374	D#	0100 10	454	F#	0111 10	534	G#	1001 10
315	A# =	1011 00	375	D# -	0100 01	455	F	0110 10	535	G	1000 10
316	A# -	1011 01	376	D#	0100 10	456	E	0101 10	536	F#	0111 10
317	A# =	1011 00	377	D# -	0100 01	457	D#	0100 10	537	F	0110 10
320	A -	1010 01	400	D	0011 10	460	D	0011 10	540	E	0101 10
321	A# -	1011 01	401	D -	0011 01	461	C#	0010 10	541	F	0110 10
322	A -	1010 01	402	D	0011 10	462	C	0001 10	542	E	0101 10
323	G# -	1001 01	403	D -	0011 01	463	B -	1100 01	543	D#	0100 10
324	A -	1010 01	404	D	0011 10	464	A# -	1011 01	544	E	0101 10
325	A# -	1011 01	405	D -	0011 01	465	D#	0100 10	545	F	0110 10
326	A -	1010 01	406	D	0011 10	466	D	0011 10	546	E	0101 10
327	G# -	1001 01	407	D -	0011 01	467	C#	0010 10	547	D#	0100 10
330	A -	1010 01	410	D#	0100 10	470	D	0011 10	550	E	0101 10
331	A# -	1011 01	411	D# -	0100 01	471	C#	0010 10	551	F	0110 10
332	A -	1010 01	412	D#	0100 10	472	C	0001 10	552	E	0101 10
333	G# -	1001 01	413	D# -	0100 01	473	B -	1100 01	553	D#	0100 10
334	A -	1010 01	414	D#	0100 10	474	A# -	1011 01	554	E	0101 10
335	A# -	1011 01	415	D# -	0100 01	475	B -	1100 01	555	F	0110 10
336	A -	1010 01	416	D#	0100 10	476	C	0001 10	556	E	0101 10
337	G# -	1001 01	417	D# -	0100 01	477	C#	0010 10	557	D#	0100 10
340	A -	1010 01	420	D	0011 10	500	D +	0011 11	560	E	0101 10
341	A# -	1011 01	421	D#	0100 10	501	C# +	0010 11	561	F	0110 10
342	B -	1100 01	422	D	0011 10	502	C +	0001 11	562	E	0101 10
343	C	0001 10	423	C#	0010 10	503	B	1100 10	563	D#	0100 10
344	C#	0010 10	424	D	0011 10	504	C +	0001 11	564	E -	0101 01
345	C	0001 10	425	D#	0100 10	505	B	1100 10	565	F -	0110 01
346	B -	1100 01	426	D	0011 10	506	A#	1011 10	566	E -	0101 01
347	A# -	1011 01	427	C#	0010 10	507	A	1010 10	567	D# -	0100 01
350	A -	1010 01	430	D	0011 10	510	A#	1011 10	570	E -	0101 01
351	A# -	1011 01	431	D#	0100 10	511	B	1100 10	571	F -	0110 01
352	B -	1100 01	432	D	0011 10	512	C +	0001 11	572	E -	0101 01
353	C	0001 10	433	C#	0010 10	513	C# +	0010 11	573	D# -	0100 01
354	C#	0010 10	434	D	0011 10	514	C +	0001 11	574	E -	0101 01
355	C	0001 10	435	D#	0100 10	515	C# +	0010 11	575	F -	0110 01
356	B -	1100 01	436	D	0011 10	516	D +	0011 11	576	E -	0101 01
357	A# -	1011 01	437	C#	0010 10	517	D# +	0100 11	577	D# -	0100 01

TABLE D6-3 (continued)

ADDR	NOTE	CODE	ADDR	NOTE	CODE	ADDR	NOTE	CODE	ADDR	NOTE	CODE
600	E -	0101 01	640	E	0101 10	700	G# -	1001 01	740	A	1010 10
601	D# -	0100 01	641	E	0101 10	701	A -	1010 01	741	s	1111 11
602	D -	0011 01	642	E	0101 10	702	A# -	1011 01	742	s	1111 11
603	C# -	0010 01	643	s	1111 11	703	B -	1100 01	743	s	1111 11
604	D -	0011 01	644	E +	0101 11	704	C	0001 10	744	s	1111 11
605	C# -	0010 01	645	s	1111 11	705	C#	0010 10	745	s	1111 11
606	C -	0001 01	646	C +	0001 11	706	D	0011 10	746	s	1111 11
607	B =	1100 00	647	s	1111 11	707	D#	0100 10	747	s	1111 11
610	C -	0001 01	650	A	1010 10	710	E	0101 10	750	A -	1010 01
611	B =	1100 00	651	s	1111 11	711	F	0110 10	751	s	1111 11
612	A# =	1011 00	652	F	0110 10	712	F#	0111 10	752	s	1111 11
613	A =	1010 00	653	s	1111 11	713	G	1000 10	753	s	1111 11
614	G# =	1001 00	654	A	1010 10	714	G#	1001 10	754	s	1111 11
615	G =	1000 00	655	s	1111 11	715	A	1010 10	755	s	1111 11
616	F# =	0111 00	656	C +	0001 11	716	A#	1011 10	756	s	1111 11
617	F =	0110 00	657	s	1111 11	717	B	1100 10	757	A =	1010 00
620	E =	0101 00	660	E +	0101 11	720	C +	0001 11	760	s	1111 11
621	E =	0101 00	661	E +	0101 11	721	C# +	0010 11	761	A =	1010 00
622	s	1111 11	662	s	1111 11	722	D +	0011 11	762	A =	1010 00
623	s	1111 11	663	s	1111 11	723	D# +	0100 11	763	A =	1010 00
624	E	0101 10	664	G# =	1001 00	724	E +	0101 11	764	A =	1010 00
625	s	1111 11	665	A =	1010 00	725	F# +	0111 11	765	A =	1010 00
626	C	0001 10	666	A# =	1011 00	726	G +	1000 11	766	A =	1010 00
627	s	1111 11	667	B =	1100 00	727	G# +	1001 11	767	A =	1010 00
630	A -	1010 01	670	C -	0001 01	730	A +	1010 11	770	s	1111 11
631	s	1111 11	671	C# -	0010 01	731	s	1111 11	771	s	1111 11
632	F -	0110 01	672	D -	0011 01	732	s	1111 11	772	s	1111 11
633	s	1111 11	673	D# -	0100 01	733	s	1111 11	773	s	1111 11
634	A -	1010 01	674	E -	0101 01	734	s	1111 11	774	s	1111 11
635	s	1111 11	675	F -	0110 01	735	s	1111 11	775	s	1111 11
636	C	0001 10	676	F# -	0111 01	736	s	1111 11	776	s	1111 11
637	s	1111 11	677	G -	1000 01	737	s	1111 11	777	s	1111 11

PARTS

In addition to parts in the standard kit you will need the following.

1 - 8-bit counter with parallel load for step counter. Suggested: 2 × 74161 or 74163. Alternate: 2 × 74191 or 74193.

1 - 9-bit up/down counter for music program counter. Suggested: 3 × 74193. Alternate: 3 × 74191.

1 - 512-word by 6-bit RAM for music program. Suggested: 3 × 2102. Alternates: 6 × 2102, 4 × 2101, 2 × 2101 (256 words).

1 - 16-word by 8-bit programmable ROM. Suggested: 1702. Alternate: 8223 or equivalent.

1 - Set of analog parts in Figure D6-2.

HINTS

The synchronous loading capability of the 74161 or 74163 makes it best suited for the step counter. The 512 by 6 memory is an odd size. The suggested way of implementing it is to use three 1K by 1 bit memories in parallel to form a 1K-word by 3-bit memory. Then two memory cycles are used to read or write a 6-bit word. Another alternative is to use six 2102's in parallel and waste half the space (unless you plan to write a 1K music program), or use wider memories such as 2101's. With two 2101's (256 by 4) you can make a 256-word by 6-bit memory. Then you will have to edit the sample program to make it shorter.

For the ROM you can use any of many field programmable ROMs. Erasable ROMs such as the 1702 are widely available, or the instructor may provide a pre-programmed ROM for the use of the entire class.

OPTIONS

Since recording music programs is the most tedious part of the synthesizer's operations, you may want to figure out ways of speeding it up. For example, modifying the ENTER switch so that it both enters a code and advances the music program counter will mean half as many push button depressions to make while programming. You will probably find it useful to have the system play the currently addressed note during recording operation.

Another possibility is to eliminate the programming features and to use programmable ROMs to hold the music program. This reduces the complexity of the project.

ACKNOWLEDGEMENTS

The programmable music synthesizer was originally designed by Moe Rubenzahl, and the arrangement of "The Flight of the Bumble Bee" is by Bruce Almich.

E-Series
8. Four-Week Projects

E1 - DIGITAL INTEGRATORS

SUMMARY

This project describes techniques for digital integration. The object of the project is to design digital integrators and to interconnect two integrators to solve a second-order differential equation.

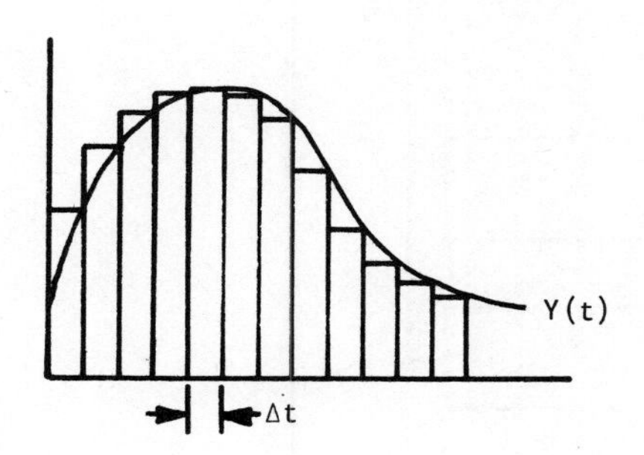

FIGURE E1-1 Digital integration.

DIGITAL INTEGRATION

As shown in Figure E1-1, an integral $\int Y(t)dt$ may be approximated by a sum of rectangle areas, $\Sigma Y(t)\Delta t$. The width of each rectangle is the iteration period, Δt, and the height of each rectangle is the value of $Y(t)$ where t is a multiple of Δt.

The iteration period Δt is the smallest quantum of time and hence it is represented as one unit digitally. Thus the area of each rectangle is numerically equal to its height. Keeping this in mind we can now describe a digital integrator. As shown in Figure E1-2, the device has two registers Y and I that contain signed quantities. The Y register holds the current value of the function $Y(t)$, and the I register accumulates the integral. During each iteration period, two operations take place.

1) The Y register is updated to hold the new value of the function $Y(t)$.

2) The new value of the Y register is added to the I register to form the current value of the integral.

Notice that the I register has twice as many bits as the Y register. Since the integral accumulates quickly when Y is large, I must be large enough to prevent overflow. The most significant n bits of I contain the usable integral with n bits of significance. This is somewhat akin to the $2n$-bit product of two n-bit numbers, in which the precision of the product is really only n bits, not $2n$ bits. The reason for choosing exactly $2n$ bits for I will become apparent later.

Practical digital integrators use an incremental representation to increase speed and simplify interconnections. Hence, instead of specifying a completely new value of Y in step (1) at each iteration period, only a ΔY increment of +1, -1, or 0 is specified. That is, the value of Y changes by at most +1 or -1 at each step. The Y register can then be a counter that at each iteration period either counts up, counts down or remains the same.

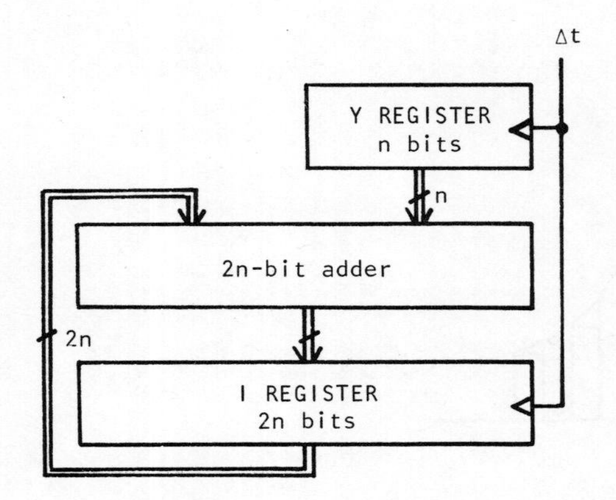

FIGURE E1-2 Digital integrator.

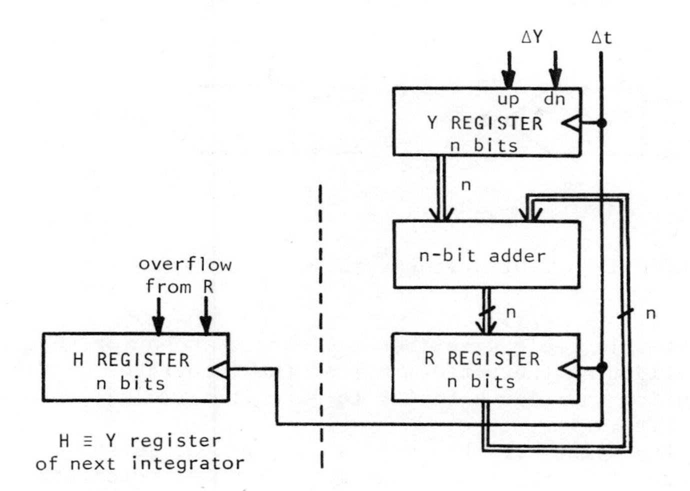

FIGURE E1-3 Incremental digital integrator.

We see that the input of a digital integrator is in an incremental form. For integrators to be interconnected to solve differential equations, the outputs must also have an incremental form. Consider the integrator of Figure E1-2 as re-drawn in Figure E1-3. The Y register has n bits and the I register is divided into an n-bit low order part R and an n-bit high order part H that contains the most significant bits and sign of the accumulated integral. Each time the Y register is added to R, the high order part can change by at most +1 or -1. Hence the overflow from the R register is the incremental output of the integrator. It indicates how much the accumulated integral changes at each step. Instead of providing an H register in each integrator, the incremental output is accumulated in the Y register of a succeeding integrator when integrators are interconnected to solve equations.

In this project you will build digital integrators that operate as described above. A single digital integrator consists of the circuitry in the right half of Figure E1-3 - an n-bit up/down counter, n-bit adder, and n-bit register - plus a small amount of combinational circuitry to generate the incremental output.

The number in the Y register will be assumed to be a two's-complement number. The incremental output that should be produced depends on the sign of Y and the carry out of the n-bit adder, as summarized in the following table:

Sign of Y	Output Carry	Incremental Output
+	0	0
+	1	+1
-	0	-1
-	1	0

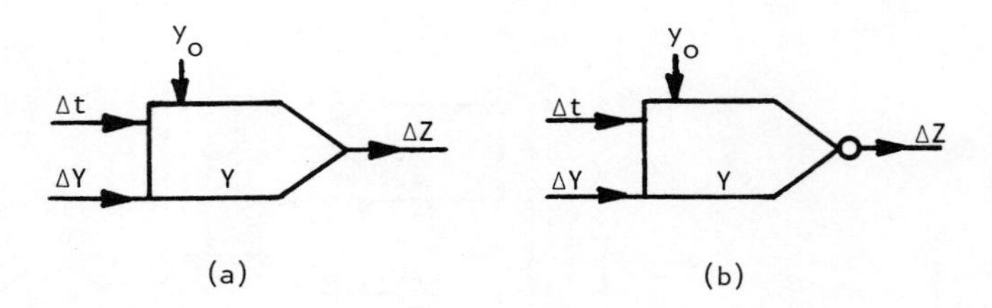

FIGURE E1-4 Integrator symbols.

That this table describes the correct behavior should be verified by considering the performance of the integrator of Figure E1-2 and extending the sign bit of Y to make it a two's-complement number with $2n$ bits.

The symbol for the digital integrator is shown in Figure E1-4(a), where Δt is the time quantum, ΔY is the incremental Y input, ΔZ is the incremental output, Y is the accumulated input, and Y_0 is the initial value of Y. It is possible to perform a sign inversion by swapping the +1 and -1 incremental signals, and Figure E1-4(b) is the symbol for such an inverting integrator. The output of an integrator is the integral of its input; equivalently, the input is the derivative of the output.

ASSIGNMENT

I. Design and construct an 8-bit integrator. Use an 8-bit up/down counter, an 8-bit adder, and an 8-bit register as the basic components. You will also need a small amount of combinational circuitry for deriving the incremental output. If 74191s are used for the up/down counter, then the format of the incremental output can be directly compatible with the 74191 count inputs. That is, it can consist of two signals that may be connected directly to the enable and up/down inputs of the counter in the next integrator. You should provide a push button and eight toggle switches for initializing the Y register and clearing the R register, an input for the Δt clock, and eight lamps for the counter output (Y). To aid you in the rest of the project, wire up a clock controller such as described in Project B1 or use a pulse burst generator.

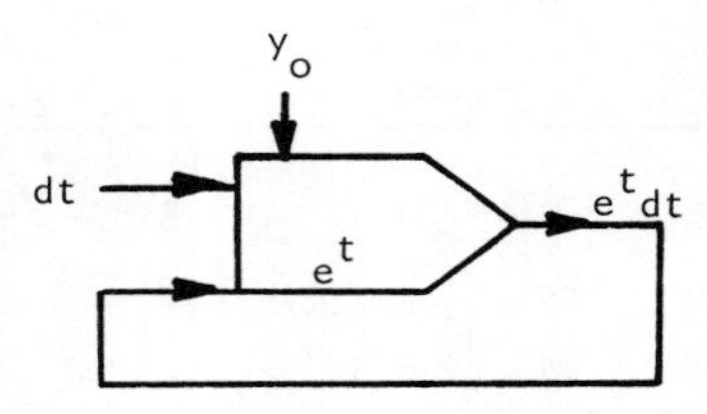

FIGURE E1-5 Exponential loop.

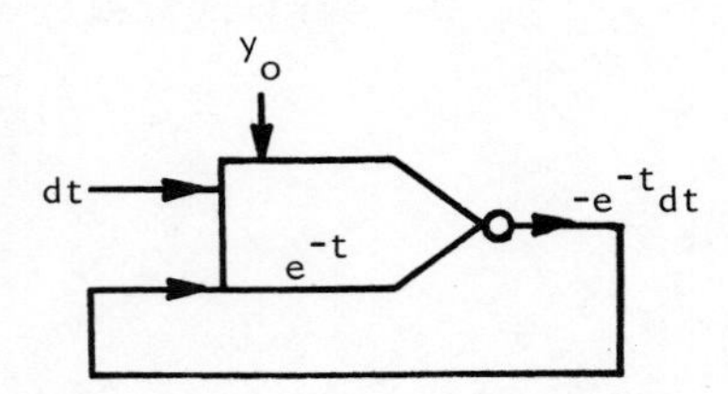

FIGURE E1-6 Negative exponential loop.

II. Hook up your integrator to solve the equation $\frac{dy}{dt} - y = 0$, as shown schematically in Figure E1-5. Ignoring scale factors the accumulated integral is e^t because $\frac{d(e^t)}{dt} = e^t$, or $d(e^t) = e^t dt$. Taking into account scale factors, the accumulated integral after N time quanta should be approximately $y_0 e^{N/256}$.

TABLE E1-1 Exponentials

N	$32e^{N/256}$	$127e^{-N/256}$
0	32.0	127.0
15	33.9	119.8
30	36.0	113.0
45	38.1	106.5
60	40.5	100.4
75	42.9	94.7
90	45.5	89.3
105	48.2	84.3
120	51.1	79.5
135	54.2	75.0
150	57.5	70.7
165	61.0	66.7
180	64.6	62.9
195	68.5	59.3
210	72.7	55.9
225	77.1	52.7
240	81.7	49.7
255	86.6	46.9
270	91.9	44.2
285	97.4	41.7
300	103.3	39.3
315	109.6	37.1
330	116.1	35.0
345	123.1	33.0
360	130.6	31.1

Using an initial condition of $y_0 = +32$, plot the output of the integrator from $t = 0$ to $t = 345$ in steps of 15. Plot the exact values of $32e^{N/256}$ given in Table E1-1 and compare.

III. Use an initial condition of $y_0 = -32$ to verify that your circuit works correctly for negative values of Y.

IV. Hook up your integrator to invert and solve the equation $\frac{dy}{dt} + y = 0$, as shown in Figure E1-6. Use the initial condition $y_0 = +127$ and plot the integrator output for $t = 0$ to $t = 345$ in steps of 15. Plot the exact values of $127e^{-N/256}$ given in Table E1-1 and compare.

V. Build an integrator that replaces the adders and registers of Figure E1-3 with a 6-bit binary rate multiplier. (Leave the existing integrator intact.) Repeat Assignments II and IV with the BRM-based integrator and compare and comment on the results.

VI. Suppose you had to build a 12-bit integrator. What ICs would you need for a BRM-based integrator and for a register/adder integrator? What are the trade-offs?

VII. Dismantle the BRM-based integrator and build another register/adder integrator. Connect the two integrators in a sine-cosine loop to solve $\frac{d^2y}{dt^2} + y = 0$ as shown in Figure E1-7.

The first integrator has an initial condition of 0 and the second has initial condition y_0, the amplitude of the sine wave output. The accumulated integral in integrator 1 after N steps is approximately $y_0 \cdot \sin(N/256)$; and in integrator 2 the approximate function $y_0 \cdot \cos(N/256)$ is accumulated.

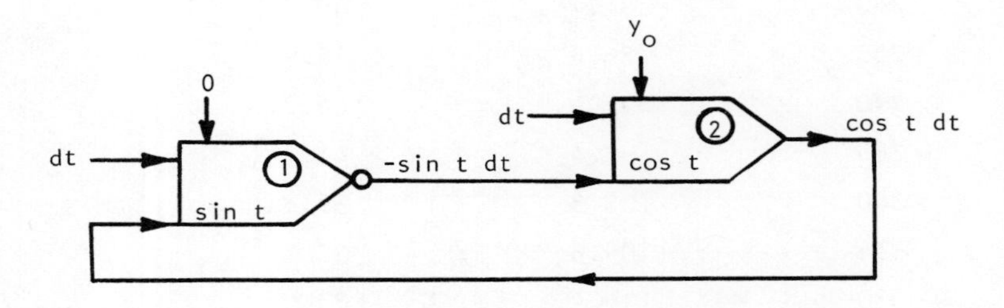

FIGURE E1-7 Sine-cosine loop.

TABLE E1-2 Sines and Cosines

N	100 sin(N/256)	100 cos(N/256)
0	0.0	100.0
30	11.7	99.3
60	23.2	97.2
90	34.4	93.9
120	45.2	89.2
150	55.3	83.3
180	64.7	76.3
210	73.1	68.2
240	80.6	59.2
270	87.0	49.3
300	92.1	38.8
330	96.1	27.8
360	98.6	16.4
390	99.9	4.7
420	99.8	-7.0
450	98.3	-18.6
480	95.4	-30.0
510	91.3	-40.9
540	85.8	-51.3
570	79.3	-61.0
600	71.6	-69.8
630	62.9	-77.7
660	53.4	-84.5
690	43.2	-90.8
720	32.3	-94.6
750	21.0	-97.8
780	9.5	-99.6
810	-2.2	-99.9

Plot the accumulated cosine integral for $t = 0$ to $t = 810$ in steps of 30, with an initial condition $y_0 = 100$. The exact values of 100 cos(N/256) and 100 sin(N/256) are given in Table E1-2. Plot the exact value of the cosine and compare.

VIII. Error analysis of the sine-cosine loop shows that the actual accumulated integrals in Assignment VII have the form

$$I_1 = k^N \sin N\alpha$$

$$I_2 = k^N \cos N\alpha$$

where $k = (1 + 1/M^2)^{\frac{1}{2}}$, $\alpha = \tan^{-1} 1/M$, and $M = 256$. Hence the accumulated integrals drift from the correct integrals in both amplitude and phase over periods of several cycles.

The solution of the equation $N/256 = 20$ is $N = 16085$, so that 16085 steps are needed for 10 cycles of the sine wave. With an initial condition of $y_0 = 100$, observe the drift for 10 cycles of the sine wave by making 10685 steps. The computed value of k^{16085} is 1.13, and hence the amplitude of the sine wave after 10 cycles should be about 113. The phase drift, on the other hand, is not observable after this few cycles.

IX. The integrator loop designed above is of the "simultaneous" variety, because all of the integrators change at the same time. A "sequential" scheme can also be used, in which one integrator changes before the other. Modify your sine-cosine loop to make it sequential by providing two clocks and updating one integrator with the first clock and the other with the second, as shown in Figure E1-8. Hence, the first integrator is updated, then the second, then the first, etc.

Error analysis of the sequential scheme indicates that all errors are either sinusoidal or bounded, so that this scheme does not exhibit the phase or amplitude drifts of the simultaneous scheme. Verify this by running the new sine-cosine loop for a long time.

X. Hook up an 8-bit DAC to the sine wave integral (invert the sign bit to get a proper signed display), and observe the DAC output on an oscilloscope. Neglecting DAC settling, what is the smallest Δt for which your system will work? What is the real-time frequency of the sine wave for this Δt?

Slow down the frequency of the system so that DAC settling is not a problem. Hook up an analog sine wave generator to the second scope channel and compare the two waveforms.

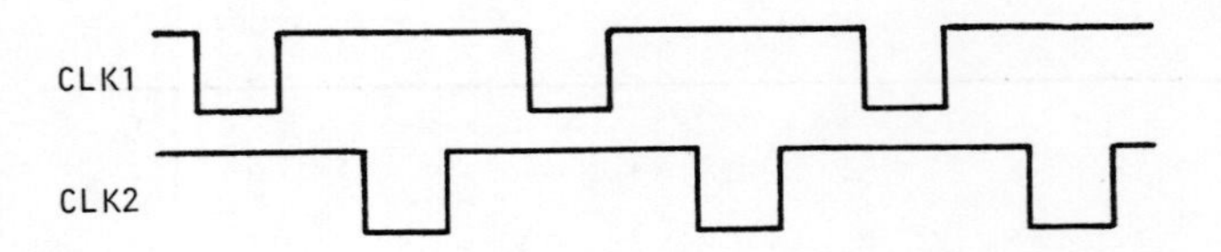

FIGURE E1-8 Two-phase clock.

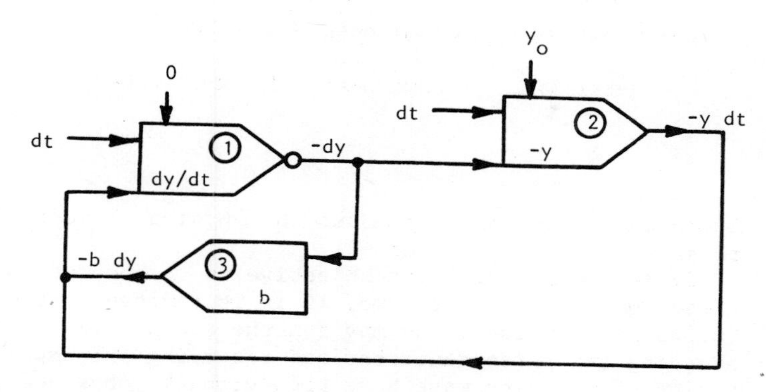

FIGURE E1-9 Damped sine-cosine loop.

XI. Modify your system to produce damped sine waves by solving the equation $\frac{d^2y}{dt^2} + b\frac{dy}{dt} + y = 0$, as shown schematically in Figure E1-9.

In Figure E1-9, integrator 3 performs multiplication by a constant. The Y register of integrator 3 is initially loaded with the constant b and does not change during the integration (the ΔY input is not used). Whenever +1 increments are produced by integrator 1, integrator 3's Y register is added its R register; when -1 increments are produced, -Y is added to R. The incremental output of integrator 3 is derived from the overflow of R in the usual manner.

Implement integrator 3 as a 4-bit integrator. Since the Y-register does not change, you may simply use 4 toggle switches or a wired connection to set the value of b.

The ΔY input of integrator 1 must respond to the outputs of both integrators 2 and 3. Therefore you will need circuitry to combine the two outputs. You may want to implement a two-phase system in which integrator 1 responds first to the output of integrator 2 and then to integrator 3. The overall system operation may be simultaneous or sequential.

Provide some circuitry for resetting the initial conditions automatically when the output amplitude becomes small. Then display the output on the oscilloscope. Observe the exponential decay and the frequency of the output for various values of b. What is the solution of the differential equation being emulated?

PARTS

In addition to parts in the standard kit, you will need the following:

2 - 8-bit up/down counters. Strongly recommended - 4 × 74191. Alternate - 4 × 74193.

2 - 8-bit adders. Recommended - 4 × 7483.

2 - 8-bit registers. Recommended - 4 × 74175 or 74174.

1 - 6-bit BRM. Recommended - 1 × 7497.

1 - 8-bit DAC. Recommended - 1 × MC1408L-8.

HINTS

Spend some time thinking about the physical layout of this project to make the wiring easier.

In Assignment VIII the objective is to observe the amplitude error after 10 cycles. Therefore, if it is inconvenient for you to count 16,085 clock pulses, you may run the clock from a pulse generator, stopping the pulse generator and recording the amplitude when 10 cycles of the sine wave have been visually observed in the lights.

You may wish to complete the visual display described in Assignment X first to aid the completion of Assignments VIII and IX.

OPTIONS

The project can be made effectively a D-series project by eliminating Assignment XI.

To extend the project, in Assignment XI hook up another 8-bit DAC to the cosine integral. Connect the cosine DAC to the X-input and the sine DAC to the Y-input of the scope in X-Y mode. A spiral waveform should be produced.

REFERENCES

Digital integration techniques are discussed by Sizer [1968].

E2 - OSCILLOSCOPE PING-PONG

SUMMARY

This project requires the design of a single-user Ping-Pong game similar to the commercially available two-user games.

INTRODUCTION

The purpose of this project is to display a moving ball on the screen of the oscilloscope, which bounces off the edges of the screen and responds to the Ping-Pong paddle of a user.

The circuit of Figure E2-1 is the most important element of the system. The binary rate multiplier (BRM) has a 6-bit delta X (DX) as its rate inputs, and the BRM output is used to enable an 8-bit counter. The counter output is connected to a digital-to-analog converter (DAC) that provides the X input voltage for the X-Y oscilloscope. An identical circuit is used for the Y input. The DX and DY values for the X and Y channels determine the direction and rate of movement of the scope trace (Ping-Pong ball). The Ping-Pong ball

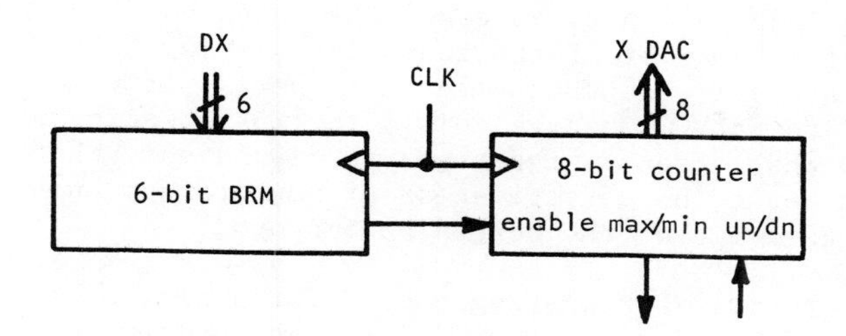

FIGURE E2-1 Ping-Pong circuit.

has reached an edge of the screen when either the X or Y counter reaches its maximum or minimum value. The ball can be "bounced" off the edge by simply reversing the counting direction of the appropriate counter.

If the DX and DY values for the two channels are constant, then the behavior of the bouncing ball will be quite predictable and will exhibit a regular pattern. The behavior can be made more interesting by automatically changing DX and DY when the ball hits certain edges. One way to do this is to use random numbers for DX and DY, getting a new random number each time the ball hits an edge. An 8-bit random number can be generated as described in Project A7. Half the bits of this number can be used for DX and the other half for DY. (The remaining 2 bits of DX and 2 bits of DY can be constant or can be hooked up in some interesting manner.)

Two copies of the circuit of Figure E2-1, control circuitry for changing counter direction, and a random number generator can easily be used to construct a system that produces an interesting "bouncing ball" on the oscilloscope screen. To play Ping-Pong, you must add a paddle and control circuitry for moving and displaying the paddle and detecting when the ball strikes the paddle.

ASSIGNMENT

The Ping-Pong game should be built and tested in stages as outlined below. At each step keep in mind that you will be adding more circuitry later -- keep your layout and wiring compact so that you do not run out of breadboard space.

I. Build and test an 8-bit random number generator.

II. Build two copies of the circuit of Figure E2-1, and hook up the counter output of each circuit to an 8-bit DAC (see Project A9). Add control circuitry for edge detection and changing the counter direction. Connect the DAC outputs to the X and Y inputs of an oscilloscope, and test the system using constants for the DX and DY inputs. Adjust the system clock so that the "ball" on the scope screen moves at a reasonable speed.

III. Connect the random number generator to the DX and DY inputs, and provide control circuitry for producing a new random number when the Ping-Pong ball hits an edge (this need not happen at all edges). Observe the behavior of the resulting display. The behavior of the system is cyclic, so that after some period of time the ball will retrace its path. Observe this effect by speeding up the system clock so that a constant pattern is displayed. What does this pattern tell you about the effectiveness of your random number generator? How can the coverage of the screen be improved?

IV. Provide a Ping-Pong paddle for a single user. The paddle should be a vertical line on the screen, and there should be two push buttons for moving the paddle up or down. You have to figure out how to display the paddle on the screen. The paddle should be 32 dots high. When the Ping-Pong ball hits the left edge of the screen, it should bounce if it hits anywhere on the paddle and disappear if it misses the paddle. Provide a means for restarting the game after a miss.

PARTS

In addition to parts in the standard kit, you will need the following integrated circuits:

2 - 8-bit DACs. Recommended: MC1408L-8.

2 - 8-bit up/down counters for X and Y. Recommended: 4 × 74191. Alternate: 4 × 74193.

2 - 7497 BRM.

1 - 8-bit parallel-out shift register for random numbers. Recommended: 1 × 74164 or 2 × any 4-bit parallel-out shift register.

1 - 8-bit up/down counter for paddle position. Recommended: 2 × 74193 or 74191. Alternate: Potentiometer for analog paddle positioning.

1 - 8-bit comparator for paddle hit. Recommended: 2 × 7485. Alternate: analog comparator such as LM311.

2 - 8-bit, 2-input multiplexers for ball/paddle display. Recommended: 4 × 74157 or 74158.

HINTS

When the paddle is drawn, the Y-position of each of the dots that comprise it should be compared with the Y-position of the ball. If the positions are equal at any point and the ball is at the edge, then the ball bounces. The comparison can be performed with either a digital or an analog comparator, depending on the paddle implementation.

A straightforward implementation of the project requires twenty-five to twenty-eight ICs, the most of any project in this manual.

However, it still is possible for a very clever designer to reduce the number of ICs so that the project will fit in three standard plug-in strips. One reduction is to eliminate X-axis multiplexers for the ball/paddle display. This may be accomplished by allowing the paddle to move horizontally on the screen with the ball instead of fixing it at one edge. Another possibility is to switch the DAC V_{ref} input between 0 and 3 volts to accomplish the paddle display. If your scope has two vertical inputs in display mode, the Y-axis multiplexers can also be eliminated at the expense of another DAC.

OPTIONS

The project can be reduced to the C-series level by omitting Assignment IV.

The game is easier to play if you include circuitry to prevent the paddle from wrapping around at the top and bottom of the screen. The up or down push button would be disabled when the paddle touches the top or bottom edge.

Many other games are possible if you use your imagination. One game that can be done by a clever designer without too much extra circuitry is called "Etch-a-Pong." It is a combination of this project and Project D3. Initially a small raster of 32 by 32 dots is placed on the screen. There is a paddle and bouncing Pong-Pong ball. Whenever the ball gets near a dot, it erases it. The object of the game is to bounce the ball around so that all of the dots are eventually erased. Instead of producing a random return when the ball hits the paddle, the circuit is designed so that the position of the paddle hit determines the angle of return. Thus a skilled player can erase the entire screen in about five minutes. "Etch-a-Pong" can be constructed with about 30 ICs and fits in four standard plug-in strips. It is 25 to 75 percent more complex than this project, depending on whether the student has done this project or Project D3 before.

E3 - FLOATING-POINT ADDER

SUMMARY

In this project a floating-point adder will be designed that takes as inputs two 12-bit, signed, possibly unnormalized numbers and ouputs a 12-bit, signed, normalized sum and an overflow indicator.

INTRODUCTION

The floating-point format used in this project is as follows:

Z	X	X	X	X	Y	Y	Y	Y	Y	Y	Y
11	10	9	8	7	6	5	4	3	2	1	0

The *mantissa* of the number is Z.YYYYYYY where Z is the sign of the mantissa and the radix point is on the left. The mantissa is in two's-complement form, so that the value of the mantissa is $M = (YYYYYYY - Z*2^7)/2^7$, and $+127/128 \geq M \geq -1$. The *exponent* of the represented number is XXXX. The exponent is in excess-8 format, so that the value of the exponent is E = XXXX - 8, and $+7 \geq E \geq -8$. The value of the represented number is $M*2^E$. A number in this format is *normalized* if the sign bit and the next most significant bit of the mantissa are different. Some examples of normalized numbers in this format are given below.

0 1111 1111111 = $(127/128)*2^7$ = 127 (largest positive number)

1 1111 0000000 = $(-128/128)*2^7$ = -128 (largest negative number)

0 0000 1000000 = $(64/128)*2^{-8}$ = 2^{-9} (smallest normalized positive number)

0 1001 1000000 = $(64/128)*2^1$ = 1

1 1000 0000000 = $(-128/128)*2^0$ = -1

Some examples of unnormalized numbers are as follows.

0 1011 0010000 = $(16/128)*2^3$ = 1

1 1100 1111000 = $(-8/128)*2^4$ = -1

0 0000 0000000 = $(0/128)*2^{-8}$ = 0 ("clean" zero)

0 1001 0000000 = $(0/128)*2^1$ = 0 (a "dirty" zero)

0 0000 0000001 = $(1/128)*2^{-8}$ = 2^{-15} (smallest positive number)

Note that there are many possible representations of zero, but only one of them is defined to be "clean." An attempt to normalize the smallest representable nonzero numbers (in the range $2^{-9} > N \geq -2^{-9}$) results in *exponent underflow*.

ASSIGNMENT

Figure E3-1 gives a suggested block diagram for an adder for the floating-point numbers described above. You are to design the adder circuit subject to the following requirements.

1) There are two push buttons, PB1 and PB2. PB1 is used to reset the system and initialize, and PB2 loads segments of the operands as follows:

   a) On the first depression, the 8-bit mantissa of the first operand is loaded from eight toggle switches.

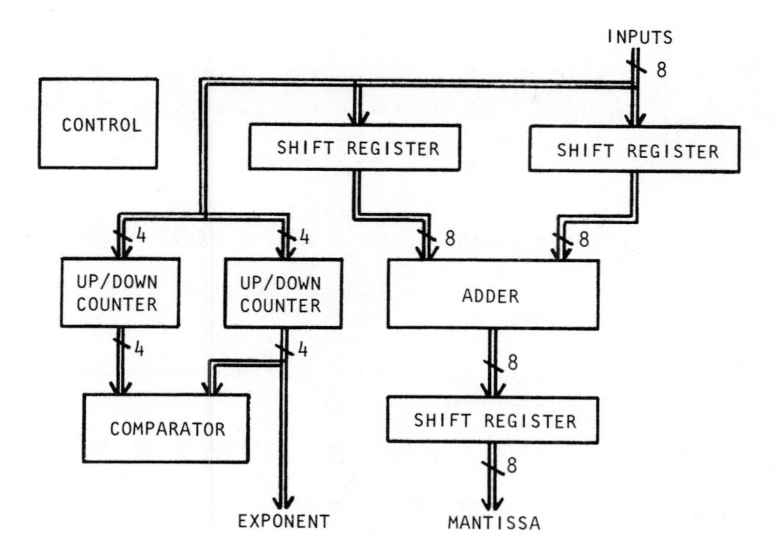

FIGURE E3-1 Block diagram.

b) On the second depression, the 8-bit mantissa of the second operand is loaded.

c) On the third depression, the two 4-bit exponents are loaded. Execution proceeds automatically and the result is displayed in 12 lamps.

2) A free running external clock (pulse generator) is used to provide timing.

3) Overflow is indicated by a lamp.

4) All true zero results are to be automatically converted into "clean zero" form.

5) All results that produce exponent underflow should be automatically converted into "clean zero" form and indicated by an underflow lamp.

6) Input operands may be unnormalized, but the result must always be normalized if it is nonzero.

Answer the following questions:

1) Are there special cases where a valid result might be obtained from a more complex floating-point adder whereas your design will produce overflow or underflow?

2) How could your design be improved to handle more special cases in a correct manner? What is a "guard digit"?

3) How could your design be made to execute faster?

4) Explain how an arithmetic comparison can be easily done between two numbers in the given floating-point representation.

PARTS

In addition to parts in the standard kit, you will need the following:

3 - 8-bit parallel-in bidirectional shift registers. Suggested: 6 × 74194.

1 - 8-bit parallel-in shift register. Suggested: 2 × any 4-bit shift register.

1 - 8-bit adder. Suggested: 2 × 7483 or 74283.

2 - 4-bit up/down counters. Recommended: 2 × 74191. Alternate: 2 × 74193.

1 - 4-bit comparator. Suggested: 1 × 7485.

HINTS

The control circuitry is obviously the most difficult part of this assignment, since the data paths are given. Do a clean design of a sequential machine to cycle through the control states needed to automatically execute your floating-point addition algorithm.

It is possible to eliminate the post-normalization shift register by suitably shifting and re-adding the input operands instead.

OPTIONS

Use a different floating-point representation, for example, sign-magnitude for the mantissa and two's-complement for the exponent.

Operand loading can be done in two steps if twelve toggle switches are available.

REFERENCES

Floating-point representations and hardware are discussed in Gschwind and McCluskey [1975], Stone [1972], Booth [1971], and most other texts on computer organization.

E4 - DIGITAL STORAGE SCOPE

SUMMARY

An ordinary oscilloscope can be used in conjunction with analog-to-digital converters, a memory, and refresh circuitry to create a storage oscilloscope that displays signals indefinitely after they disappear. This project gives only the rough idea of such a system and leaves most of the system design to the student.

INTRODUCTION

Storage oscilloscopes are useful for observing one-time events and slowly changing signals that do not produce a readable trace on a conventional oscilloscope. Conventional storage scopes use special phosphors that retain a trace after it is written. A digital storage scope converts the input signal to a sequence of digital values, stores this sequence in a memory, and refreshes the screen of a conventional scope from the memory.

ASSIGNMENT

Design a digital storage scope that displays input signals in a 256 by 256 dot matrix. Analog input signals should be converted to 8-bit digital values at a rate determined by an adjustable sweep clock. These 8-bit values should be stored in sequential locations of a 256 word by 8 bit RAM. There should be an independent refresh clock that fetches values from the RAM and displays then on an X-Y scope such that a viewable trace is produced on the X-Y scope, say one sweep every 1/60 second. The RAM address is used as the input to an X DAC and the RAM output as the input to a Y DAC for the X-Y display.

PARTS

You will need a 256 by 8 RAM, an 8-bit analog-to-digital converter, two 8-bit DACs, some counters, and of course the standard kit.

OPTIONS

The system described above has no triggering -- the sweep runs continuously. Design a digital triggering circuitry to add single sweep and normal sweep to the scope's capabilities. The circuit should be capable of triggering on either positive or negative edges, and should have adjustable sensitivity. Additional triggering options can be as fancy as you like.

E5 - TURING MACHINE

SUMMARY

A Turing machine is an abstract model of computation. Although Turing machines are very primitive, it has been conjectured that everything that is computable is computable on a Turing machine. In this project you will design and program a Turing machine simulator.

TURING MACHINES

Turing machines take their name from A. M. Turing, who first introduced them in a paper dealing with the theory of computation. The elements of a Turing machine are a control unit, a read/write head, and a scratch tape. The scratch tape is an infinitely long linear tape divided along its length into cells; each cell can store either a 0 or a 1. At any instant a particular cell is positioned under the read/write head, which has the ability to read or write that cell under the direction of the control unit. The head can be moved to the left or right one cell at a time along the tape, under the direction of the control unit.

The computation performed by a Turing machine is determined by the control unit, the initial contents of the tape, and the initial position of the head. A different control unit is used for each different Turing machine computation. The control unit is a Mealy model sequential machine with n states.* This machine has a single input bit, the contents of the tape cell positioned under the head. The sequential machine has two output bits -- a new bit to be written at the current head position and a bit that indicates whether the tape is to be moved to the left or to the right after writing. Hence each step of a Turing machine computation can be described as follows.

1) Read the symbol positioned under the head.

2) As a function of the current state and symbol just read, determine the new symbol to write, the direction to move, and the next state.

3) Write the new symbol.

4) Move the head.

5) Go to the next state.

The structure of a Turing machine is illustrated in Figure E5-1.

---

*In a *Mealy model*, the output depends on the current state and input. In a *Moore model* sequential machine, the output depends on the current state only.

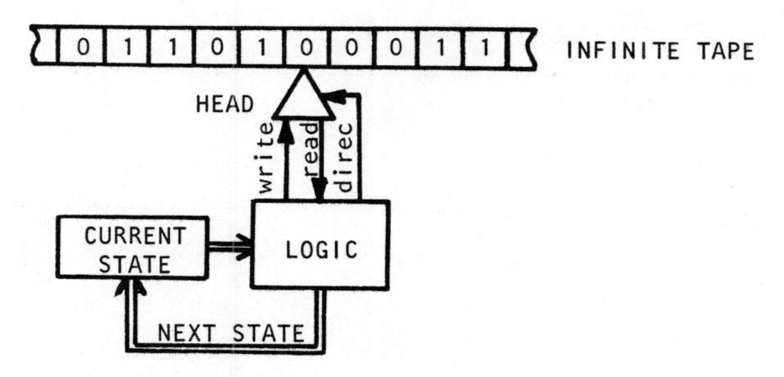

FIGURE E5-1 Turing machine.

The state table of the sequential machine that controls a Turing machine can be thought of as a *Turing machine program*. For example, suppose that we represent a positive integer i on the tape as a string of i 1's bounded by 0's. The following simple program adds one to such an integer when started in state 1 with the head positioned on the leftmost 1 of the input integer.

Current State	Current Input	Next State	Write Bit	Direction	Comments
1	0	HALT	1	R	Write a 1 and halt
1	1	1	1	R	Scan right over 1's

Notice that we have introduced a special state called HALT. In this special state the computation ceases, indicating that the program is terminated and the result can be found on the tape. The "increment" program above operates by scanning to the right over the i 1's representing the integer i. When it finds the first 0, it replaces it with a 1 and halts, leaving the representation of i+1 to the left of the head.

A less trivial Turing machine program is given in Table E5-1. Using the same input representation as the previous example, and starting with the head positioned on the leftmost 1, this program doubles the input integer. It does so by an iterative process. At each iteration the program scans for the rightmost 1 of the input integer, erases it, and writes two 1's in a scratch area to the left of the input integer. By doing this for each 1 of the input number i, it eventually erases the input number and builds up the number $2i$ in the scratch area. The operation of this program for an input number of 3 is illustrated in Figure E5-2. The figure shows the tape contents, head position, and state at just those moments when the head is about to change direction.

Another example is given in Table E5-2. This program compares two numbers for equality. The two numbers i and j are represented by strings of i 1's and j 1's, respectively, separated by a single 0.

TABLE E5-1 Doubling Program

Current State	Current Input	Next State	Write Bit	Direction	Comments
1	0	HALT	0	L	Done
1	1	2	1	R	Found a 1, scan right
2	0	3	0	L	Found a 0, back up
2	1	2	1	R	Scan right over 1's
3	0	-	-	-	Impossible
3	1	4	0	L	Erase rightmost 1
4	0	5	0	L	At center continue left
4	1	4	1	L	Scan left over 1's
5	0	6	1	L	Write a 1
5	1	5	1	L	Scan left over 1's
6	0	7	1	R	Write another 1
6	1	-	-	-	Impossible
7	0	1	0	R	At center, any 1's left?
7	1	7	1	R	Scan right over 1's

The program starts in state 1 with the head positioned over the 0 separating the two input numbers. It eventually halts with the head positioned over a 0 if the two numbers were equal and over a 1 if they were unequal. It performs the computation by an iterative process. At each iteration it erases rightmost 1 of the right hand input number and attempts to erase the leftmost 1 of the left hand input number. If there is a mismatch, then the program halts with the head pointing to the mismatching 1, otherwise it halts pointing to a zero.

ASSIGNMENT

You are to design and build a Turing machine simulator. Your simulator will be limited in that it obviously will not have an infinite tape, and it will only be able to accommodate programs with a certain maximum number of states.

The tape should be simulated using a 1K RAM, a 10-bit counter, and a 9-bit shift register as shown in Figure E5-3. The idea is to store most of the "tape" in the RAM, and to display only that part of the tape which is under or near the head. Instead of moving the head over the tape, we simulate moving the tape under the head. The head is assumed to be positioned over the cell marked "head cell" in

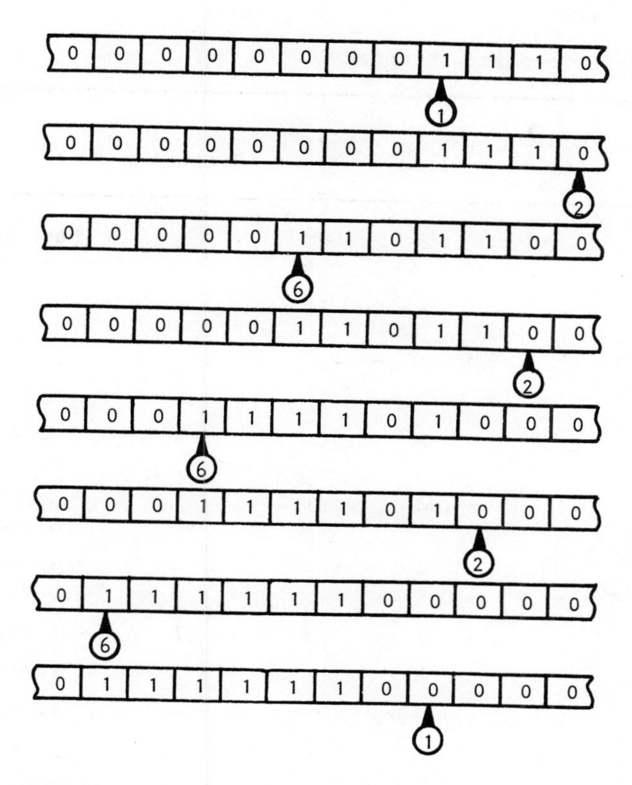

FIGURE E5-2 Doubling program execution.

Figure E5-3. The four cells on eitherside of the head are stored in parallel-out, bidirectional shift registers. Hence these nine cells in the vicinity of the head can be observed at all times with nine lamps. The other cells of the tape are stored in the RAM. A 10-bit "position counter" points to some location in the RAM. This single location is unused, but the higher locations in the RAM are assumed to contain the part of the tape to the left of the displayed cells and the lower locations are assumed to contain the part to the right.

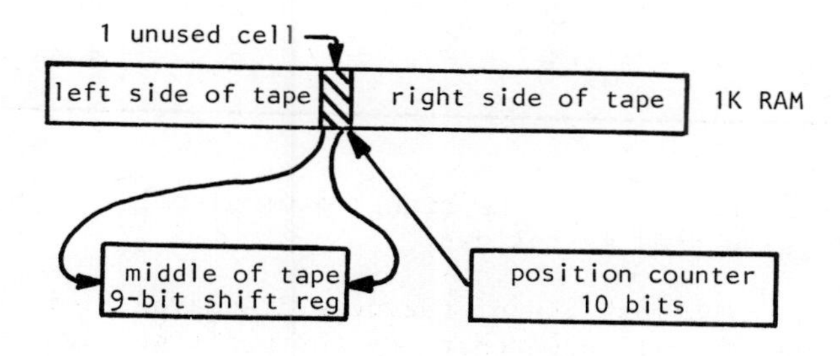

FIGURE E5-3 Tape emulation.

TABLE E5-2 Comparing Program

Current State	Current Input	Next State	Write Bit	Direction	Comments
1	0	2	0	R	Move to right of center 0
1	1	–	–	–	Impossible state/input pair
2	0	3	0	L	No 1's to right of center, look left
2	1	4	1	R	Found a 1 on right, scan right
3	0	HALT	0	L	Move left and point to answer
3	1	–	–	–	Impossible, pointing to center 0
4	0	5	0	L	At end, back up
4	1	4	1	R	Scan right over 1's
5	0	–	–	–	Impossible
5	1	6	0	L	Erase rightmost 1
6	0	7	0	L	At center, check left integer
6	1	6	1	L	Scan left to center
7	0	3	1	R	Needed a 1, mark unequal
7	1	8	1	L	Scan to leftmost 1
8	0	9	0	R	At end, back up
8	1	8	1	L	Scan to leftmost 1
9	0	–	–	–	Impossible
9	1	10	0	R	Erase to leftmost 1
10	0	2	0	R	Go back to start
10	1	10	1	R	Scan right to center

Movement of the tape to the right (equivalently, of the head to the left) is simulated as follows.

1) The rightmost bit of the nine displayed bits is written into the RAM cell pointed to by the position counter.

2) The position counter is incremented (pointing more to the left).

3) The nine displayed bits are shifted one position to the right. The rightmost bit is discarded, and the new leftmost bit is read from the RAM cell pointed to by the position counter.

Movement of the tape to the left (or of the head to the right) is similar:

1) The leftmost bit of the nine displayed bits is written into the RAM cell pointed to by the position counter.

2) The position counter is decremented (pointing more to the right).

3) The nine displayed bits are shifted one position to the left. The leftmost bit is discarded, and the new rightmost bit is read from the RAM cell pointed to by the position counter.

What remains is to simulate the Turing machine control unit, which is basically a Mealy model sequential machine. In this assignment we will be able to simulate any Turing machine with up to 16 states (including the HALT state) by using a 32 word by 6 bit RAM to store the state/output table of the control unit machine (that is, the Turing machine program). For each state, there will be two words in the RAM, one for each symbol that can be read by the head when the machine is in that state. Each word in the RAM has six bits -- four give the next state, one gives the new bit to write on the tape, and one gives the direction to move the tape.

A block diagram of the complete system is given in Figure E5-4. Your system should have the following inputs and outputs.

LEFT - A push button for manually moving the tape one position to the left.

RIGHT - A push button for manually moving the tape one position to the right.

TAPE LOAD - A push button for manually loading the head cell from a toggle switch.

TAPE IN - A toggle switch used above.

STATE LOAD - A push button for loading the 4-bit control machine state register from toggle switches.

STATE IN - Four toggle switches used above.

PROG LOAD - A push button for entering a six-bit Turing machine program word into the 32 × 6 RAM.

RUN/LOAD - A toggle switch to allow the Turing machine to operate after program and tape have been loaded.

TAPE - Nine lamps that show the contents of the head cell and the four cells on either side of it.

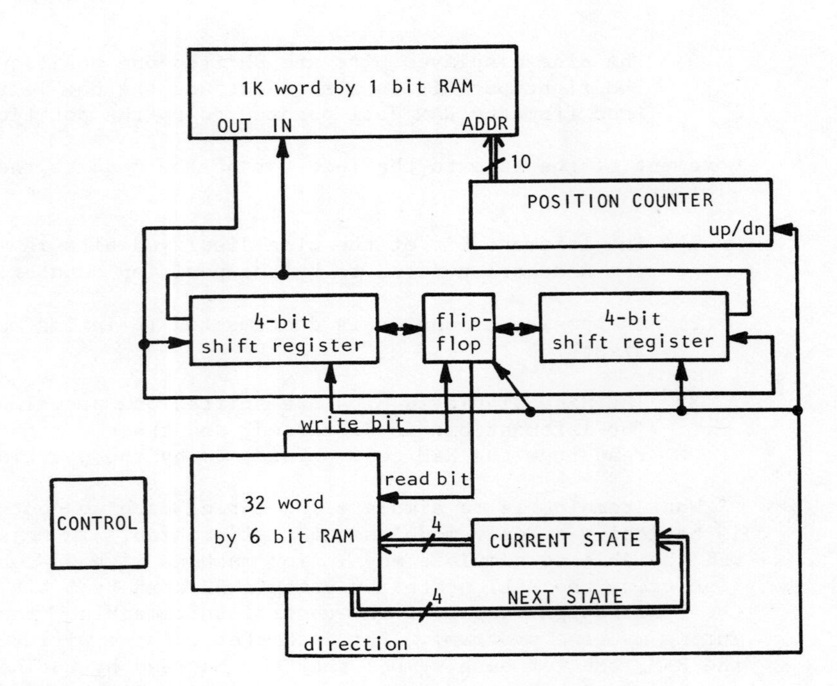

FIGURE E5-4 Turing machine block diagram.

STEP - A lamp that blinks at every Turing machine step. This enables the user to see the step even when the nine observed tape cells are the same before and after a shift.

STATE - (optional) Four lamps for displaying the current state.

PROG - (optional) Six lamps for displaying the output of the 32×6 RAM.

Your system should have a free-running clock input that controls the speed of operation of the system. The clock should be slow enough to allow the user to follow the machine's operations. The system should be tested by loading each of the example programs given earlier into the state machine and testing them with various inputs on the tape, such as 0, 1, 4, 10. The tape can be initialized manually by moving the tape and writing the cells, one cell at a time. It is also possible to write a very simple Turing machine program to reset the entire tape to zero, taking advantage of the fact that the tape is not infinite, but really wraps around.

Finally, write a Turing machine program that compacts disjoint sequences of 1's into one continuous sequence. Assume that the sequences to be compacted lie to the right of the initial head position, that they are separated by a single zero, and that at least two 0's lie to the right of the rightmost sequence. Test your program on the simulator using the input tape ...001101011100...and show that it produces...0011111100....

PARTS

In addition to the parts in the standard kit you will need the following:

1 - 1K static RAM. Suggested: 2102.

2 - 4-bit parallel-out bidirectional shift registers. Suggested: 74194. Alternate: 7495, 74195.

1 - 10-bit up/down position counter. Suggested: 3 × 74191. Alternate: 3 × 74193.

1 - 4-bit state register. Suggested: 74174 or 74175. Alternate: 74161 or 74163.

1 - 32 word × 6 bit RAM for program. Suggested 4 × 7489.

HINTS

The initial value of the tape position counter is of no importance to the simulator's operation. The simulator will operate correctly as long as the program does not use more than 1023 tape cells.

In order to properly simulate the HALT state, you may recognize one of the 16 states (say 1111) as the HALT state using a gate and use this signal to disable further operation. Alternatively, you can use an extra bit of RAM to indicate the HALT condition, creating a 32 × 7 RAM with a bit in each program step to indicate HALT or not.

Notice that the design of the Turing machine control unit is very similar to the universal sequential machine (Project C5).

OPTIONS

In order to conserve space, you may want to use a smaller RAM and counter for the tape simulation. However, the RAM should be at least 16 bits to allow some nontrivial simulations to be done. You might also want to have only a 4-bit or 8-bit tape display, and do away with the extra flip-flop (the head cell) by using the parallel-load inputs on the shift register.

To make the programming and loading of the simulator easier, you may use a counter for the state register, and you may make it possible to load all of the displayed cells of the tape at once, instead of just the head cell.

REFERENCES

Turing machines were introduced in a paper by A. M. Turing [1936]. A tutorial treatment of Turing machines, upon which this project is based, can be found in Stone [1972]. More advanced treatments of Turing machines can be found in texts dealing with the theory of computation, such as Hopcroft and Ullman [1969], Minsky [1967].

9. Microprocessor Projects

M1 - INTRODUCTION TO THE 8080

SUMMARY

This project consists of a series of assignments that introduce the student to the hardware and software features of the 8080 microprocessor.

ASSIGNMENT

I. Write a program for a small 8080 system. The system should have the features shown in Figure M1-1. The reset push button initiates an 8080 reset. The interrupt push button initiates an 8080 interrupt; the interrupt instruction executed is RST 7. The system has 256 bytes of PROM, but no RAM at all. There is a single latched output port, number 128 (80H); the output byte is displayed in eight lamps.

You should write a short assembly language program that does something interesting with the push buttons and lamps. You may come up with your own idea or develop one of the ideas below. In any case, your program must use no RAM and it must fit in no more than 256 bytes of PROM.

Note that the lack of RAM means that there is no stack for saving the processor state on interrupts. Your program should be written so that an interrupt can only take place when the program is in a known

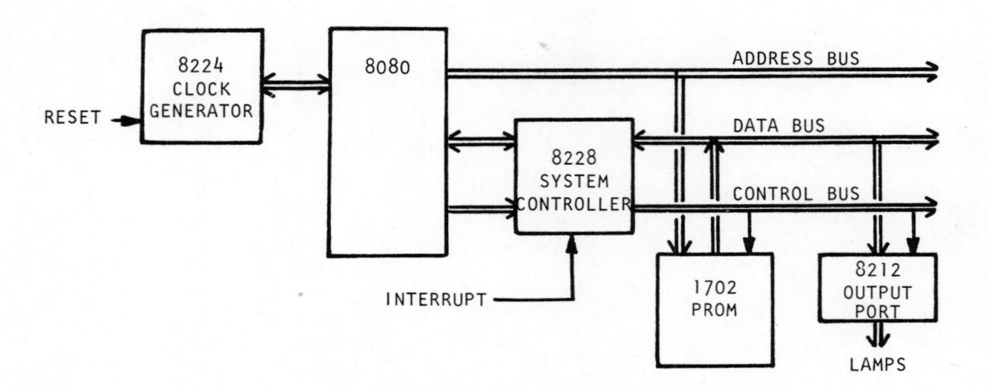

FIGURE M1-1 Small 8080 system.

state. When the interrupt processing is completed, a jump should be used rather than a return, since there is no return address on a stack.

Here are some program ideas:

1) Display zero in the output port on reset, and increment the displayed value by one each time the interrupt push button is operated.

2) Rotate a pattern in the output lamps at a speed that can be seen by the user. Complement the least significant bit of the pattern each time interrupt push button is operated.

3) Write a single player Ping-Pong game, similar to Project M4. Use the interrupt push button as the paddle.

Run and test your program on a simulator or 8080 prototyping system.

II. Explain the operation of the circuit shown in Figure M1-2. Sketch the waveforms on Ø1, Ø2, SYNC, $\overline{WR}$, and DBIN for one *instruction* cycle. Next, suppose the RDYIN input is connected to the WAIT output instead of logic 1. Explain the effect on the circuit's operation. Sketch the resulting waveforms on Ø1, Ø2, DBIN, and WAIT (=READY) for the FETCH *machine* cycle.

III. Wire up the circuit of Figure M1-2 in the lab. Do your wiring neatly to make system expansion and debugging easy. Check all voltages produced by your supplies (+12, +5, -5) before connecting power to the 8080. Also, have the teaching assistant or another student check your wiring before you turn on the power supply. If you blow out an expensive chip through carelessness, you will most likely be charged for it.

Test your circuit and turn in a sketch of the waveforms you observe on Ø1, SYNC, $\overline{WR}$, and DBIN for one instruction cycle. On the same sketch, indicate in binary the value of the data lines (D7-D0) each time SYNC, $\overline{WR}$, or DBIN is active. Are the values what you expect from the 8080 specifications? Also, observe and explain the behavior of the address lines.

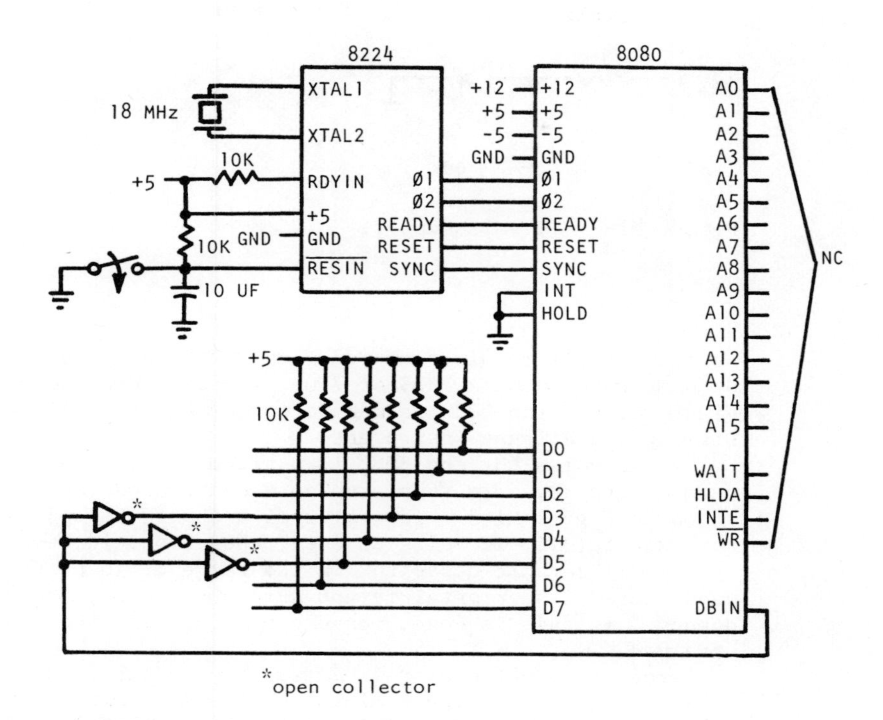

FIGURE M1-2 8080 test circuit.

Connect the WAIT output of the 8080 to the RDYIN line of the 8224 and observe and explain the effect on system behavior.

IV. Design circuitry to give an 8080 system single-step capability using the READY input. Your circuit should have a switch to place the system in single-step mode and a push button to cause the execution of a single machine cycle. Your circuit should hold READY low between steps, so that the processor is in WAIT state and the values of the address, data, and control buses can be observed.

V. Design and construct a small 8080 system that will run the program you wrote in Assignment I. Since this is a minimal system you need not decode the high-order address bits (A8-A15). The PROM should respond to any memory read. In later projects you may add RAM and circuits for address decoding.

Since your system has only one output port, you need not decode port addresses. If you choose to decode addresses, assume your system will never have more than eight output ports. Therefore you can assign the output port numbers to be 1H, 2H, 4H, 8H, 10H, 20H, 40H, 80H, so that only one bit is examined to decode the port number.

The general design of your system is of course very simple. (Refer to the 8080 system Users Manual [Intel, 1975].) The interesting aspect is making sure that all timing and fanout constraints of the 8080 and other devices are met.

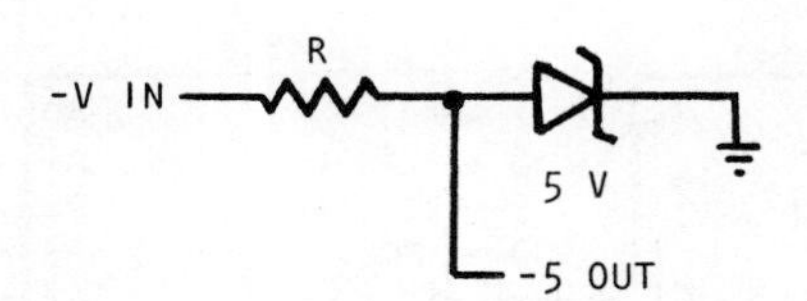

FIGURE M1-3 -5 volt supply.

PARTS

Assignment III requires an 8080, an 8224, an open-collector inverter or NAND gate such as 7403 or 7406, and discrete components shown in Figure M1-1. The 8080 is manufactured by Intel, Advanced Micro Division, and Texas Instruments.

The -5 volts for the 8080 in Assignments III and IV can be derived from a more negative supply using a zener diode and a resistor, as shown in Figure M1-3. The value of R should be approximately 50(V-5). For example, if -12 volts is used, then R should be about 350 ohms.

A small ceramic capacitor in the range of 10 pf to 20 pf may be substituted if a crystal is not available to set the 8224 clock frequency. If this is done, measure and indicate the clock frequency obtained.

Three-state buffers or inverters may be substituted for the open-collector gates in Figure M1-1. DBIN should enable the three-state outputs, and the buffer or inverter inputs should be fixed to produce 0 outputs when the outputs are enabled. Another alternative is to use an 8-bit three-state buffer (such as an 8212 with MD held at 0 and STB held at 1) to place the instruction 11000111 on the data bus when DBIN is active. In this case, the 10K pull-up resistors are not needed.

For Assignment IV you will need an 8080, an 8224 clock generator, an 8228 system controller, an 8212 or similar 8-bit latch to use as an output port, and a 1702 PROM for your program. If your breadboard does not have lamps, you can drive discrete LEDs directly from the 8212 outputs. You will also need a flip-flop such as a 7474 to detect the leading edge of the interrupt push button signal. (The level output of a push button should not be connected directly to the 8080 INT input since it may continue to request an interrupt long after the interrupt has been serviced.)

The -9 volts for a 1702 PROM can be obtained from a more negative voltage using a scheme similar to Figure M1-3. Alternatively, a series zener diode can be used, as shown in Figure M1-4 for getting

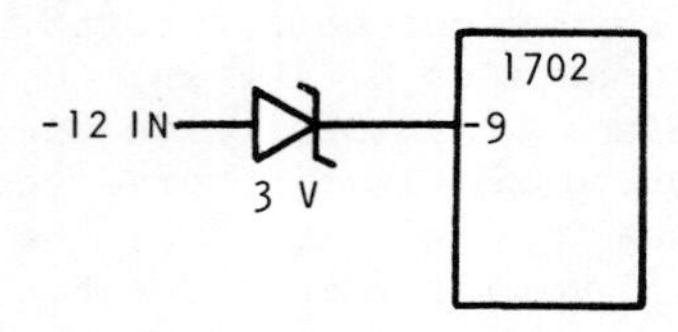

FIGURE M1-4 -9 volt supply.

-9 volts from a -12 volt supply. In this example a separate diode should be used for each 1702 in the system. (Consult the specifications for the diode and the 1702 -9 volt input current to find out why.)

M2 - INTRODUCTION TO THE 6800

SUMMARY

This project consists of a series of assignments that introduce the student to the hardware and software features of the 6800 microprocessor.

ASSIGNMENT

I. Write a program for a small 6800 system. The system should have the structure shown in Figure M2-1. The system has 256 bytes of PROM and 128 bytes of RAM. The reset push button initiates a 6800 reset. The input switches are scanned and debounced if necessary under program control. There are a number of output lamps; the number of input switches and output lamps depends on the program you choose to write. A single peripheral interface adapter (PIA) is used, so the total number of switches and lamps should be 16 or less.

You should write a short assembly language program that examines the switches for inputs and produces outputs in the lamps. You may use your own idea or develop one of the program ideas below.

1) Display a zero byte in eight output lamps on reset, and increment the displayed value each time one switch is operated, decrement each time a second switch is operated. Be sure that program works correctly if both switches are operated simultaneously.

2) Build a reaction timer. After a reset, wait a random length of time (two to ten seconds) and then flash a signal to the

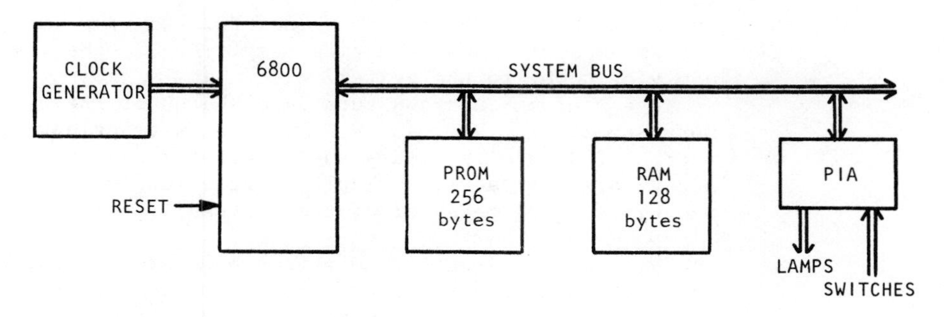

FIGURE M2-1 Small 6800 system.

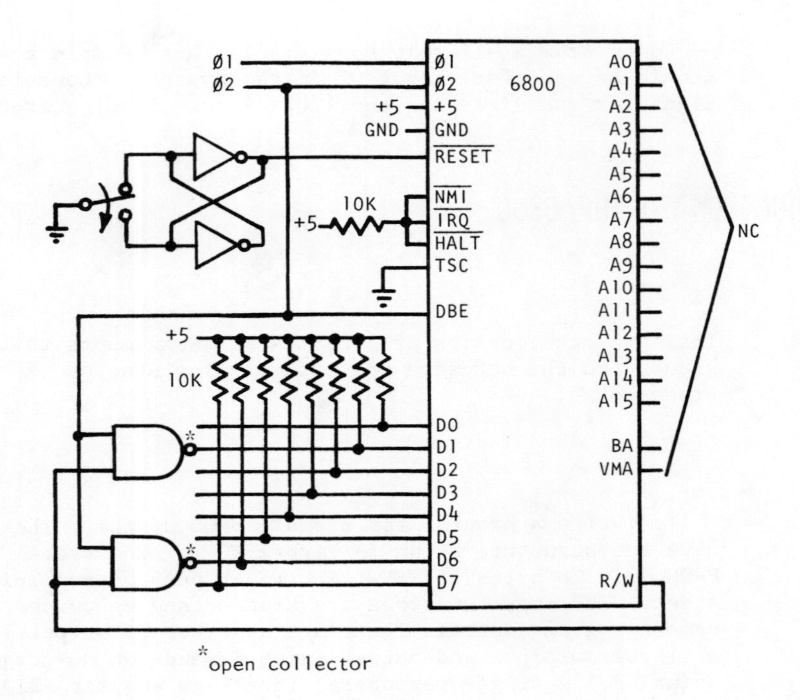

FIGURE M2-2 6800 test circuit.

user. The user is to push a button as soon as he sees the signal. Display the reaction time of the user in hundreths of a second, using two 4-bit BCD digits.

3) Implement the Ping-Pong game of Project M4.

Run and test your program on a simulator or 6800 prototyping system.

II. Explain the operation of the circuit shown in Figure M2-2. Sketch the waveforms on Ø1, Ø2, VMA, and R/W for one instruction cycle.

III. Wire up the circuit of Figure M2-2 in the lab. You will have to design your own clock generator circuit (refer to [Motorola, 1975]) or use an MC6875 to provide Ø1 and Ø2. Do your wiring neatly to make system expansion and debugging easy. Double-check your wiring before applying power to the system.

Test your circuit and turn in a sketch of the waveforms you observe on Ø1, Ø2, VMA, and R/W for one instruction cycle. On the same sketch, indicate in binary the value of the data lines each time VMA and Ø2 are both active. Explain the values you observe. Also observe and explain the behavior of the address lines.

IV. Design circuitry to give a 6800 system single-step capability, using the HALT input. Your circuit should have a switch to put the system in single-step mode and a push button to cause the execution of a single instruction in this mode. You may also wish to provide latches to display the last value of the address, data, and/or control buses during single stepping.

V. Design and construct a small 6800 system that will run the program you wrote in Assignment I. Since this is a minimal system, you need not completely decode the high-order address bits (A8-A15). Rather, use one high-order bit to select the PROM, one for RAM, and one for the PIA. In later projects you may add circuits for complete address decoding.

PARTS

In Assignment III you will need the parts shown in Figure M2-2 plus a clock generator (either an MC6875 or SSI and discrete components). The 6800 is manufactured by Motorola and American Microsystems, Inc. For Assignment V you will need in addition a 256-byte PROM such as a 1702, a 128-byte RAM such as a 6810, and a PIA such as a 6820. A small amount of SSI will be needed if you implement the circuit designed in Assignment IV.

M3 - INTRODUCTION TO THE 2650

SUMMARY

This project consists of a series of assignments that introduce the student to the hardware and software features of the 2650 microprocessor.

ASSIGNMENT

I. Write a program for a small 2650 system. The system should have 256 bytes of PROM but no RAM. It should have a number of input switches and output lamps; the number depends on the program you choose to write. One input switch can be connected to the microprocessor's SENSE input; additional switches require an input port such as an 8T31 or 8212. Eight output lamps can be connected to the system through an output port such as an 8T31 or 8212. Alternatively, a single PIA can be used for both input and output.

You should write a short assembly language program that examines the switches for inputs and produces outputs in the lamps. You may use your own program idea or develop one of the ideas given in Projects M1 and M2. Run and test your program on a simulator or 2650 prototyping system.

II. Explain the operation of the circuit shown in Figure M3-1. Sketch the waveforms on CLOCK, OPREQ, $M/\overline{IO}$, $\overline{R}/W$, and WRP for one instruction cycle.

III. Wire up the circuit of Figure M3-1 in the lab. Use a pulse

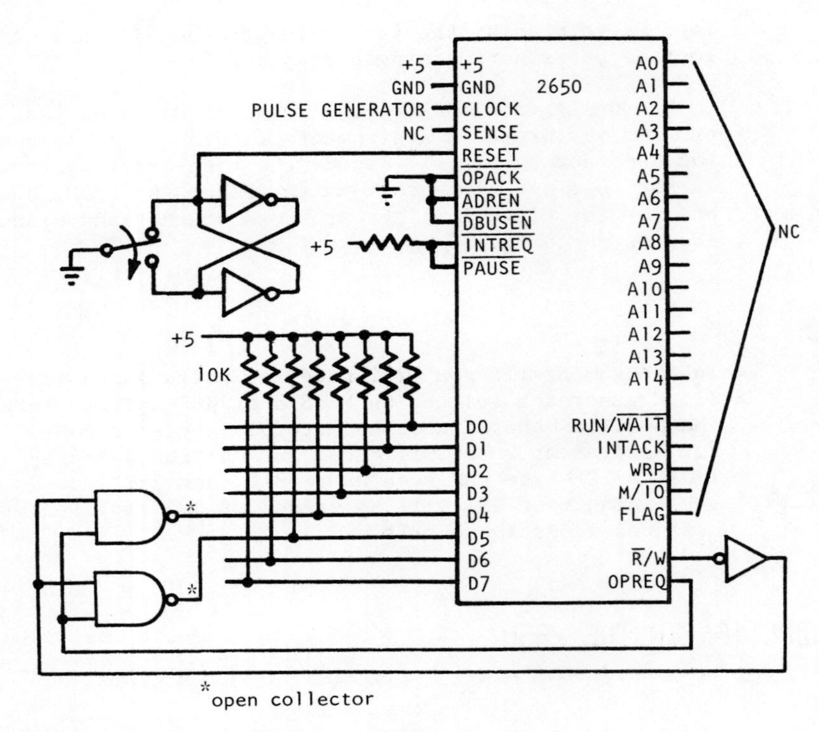

FIGURE M3-1 2650 test circuit.

generator to provide the clock input. Do your wiring neatly to make system expansion and debugging easy. Double-check your wiring before applying power to the system.

Test your circuit and turn in a sketch of the waveforms you observe on CLOCK, OPREQ, M/IO, R/W, and WRP for one instruction cycle. On the same sketch, indicate in binary the value of the data lines each time OPREQ is valid and explain the values you observe. Also observe and explain the behavior of the address lines.

IV. Design circuitry to give a 2650 system single-step capability. Your circuit should have a switch to place the system in single-step mode and a push button to cause the execution of a single machine cycle (three clock periods) in this mode. Your circuit should hold CLOCK high during T2 of a machine cycle so that control, data, and address lines may be observed after each single step.

V. Design and construct a small 2650 system that will run the program you wrote in Assignment I.

PARTS

In Assignment III you will need the parts shown in Figure M3-1. The 2650 is manufactured by Signetics and Advanced Memory Systems. For

Assignment V you will need in addition a 256-byte PROM such as a 1702 and two 8-bit I/O ports such as 8T31s or a PIA such as a 6820. A small amount of SSI will be needed if you implement the circuit designed in Assignment IV.

M4 - LAMP PING-PONG

SUMMARY

This project is a very simple microprocessor application using two input switches and an 8-bit output port.

ASSIGNMENT

You are to design and construct a microprocessor-based version of Project B2. Use eight lamps to display the ball and two push buttons for paddles. The push buttons may be scanned continuously under program control, or they may be hard-wired to produce interrupts when operated.

Microprocessor control makes a number of features very easy to implement. When a player misses the ball, the scores of both players should be flashed in the lights before the game is restarted. When a player hits the ball, the speed of the ball return should be adjusted according to how "well" he hits it. For example, a fast return can be produced if the ball is hit at the last possible moment. When the microprocessor is reset, both players' scores should be set to zero.

PARTS

You will need a microprocessor, basic system support chips, and a 256-byte PROM. With some microprocessors you will be able to keep the entire system state in registers, eliminating the need for RAM. Otherwise you will need one 128-byte RAM chip.

Two input switches are needed for the paddles, and an 8-bit output port and lamps are needed to display the ball. All I/O can be accomplished with a single PIA, or with an 8-bit output latch and two bits of input buffers.

M5 - ANALOG-TO-DIGITAL CONVERTER

SUMMARY

Analog-to-digital conversion is performed by a microprocessor system that sends trial values to a DAC and senses the output of an analog comparator.

ASSIGNMENT

Project A10 described the basic structure of an analog-to-digital converter. You are to design an analog-to-digital converter in which the block labeled "CONTROL" in Figure A10-3 is a microprocessor. Your system should have an input port to sense the output of an analog comparator and an 8-bit output port to load a DAC with a trial conversion value. Each time the microprocessor is reset, it should execute a successive approximation algorithm to determine the value of the analog input voltage. The 8-bit output port can be connected to lamps as well as the DAC so that the converted value can be observed by the user. How long does it take for a conversion in your system?

You may wish to add a number of features to your system. For example, by inserting a resistive divider network and analog multiplexer between the analog input and the comparator, you can program your system to be an auto-ranging digital voltmeter. Another possibility is to program your system to make continuous conversions using a tracking algorithm.

PARTS

You will need a microprocessor, basic system support chips, and a 256-byte PROM. With most microprocessors the entire system state can be kept in registers, so no RAM is needed.

In a minimum configuration, the only input besides the reset push button is the state of the analog comparator. Eight bits of output are needed to drive the DAC and output lamps. Obviously all I/O can be accomplished with a single PIA. Possible DACs and comparators to use are mentioned in Project A10. Some DACs such as Analog Devices AD7522 have input latches and can be placed on a microprocessor bus directly.

M6 - PROGRAMMED TERMINAL INPUT/OUTPUT

SUMMARY

In this project the serial I/O bit stream of a terminal is processed by a microprocessor program.

SERIAL COMMUNICATION

Before giving the assignment we will discuss the asynchronous serial communication techniques most commonly used by computers to communicate with terminals. There are generally two serial data streams between computer and terminal, one for transmission in each direction. The rate of transmission (bits per second) is called the *baud rate*. The inverse of the baud rate is called the *bit time*, the length of time for the transmission of a single bit.

Transmission rates for typical computer terminals are between 110 baud and 9600 baud. A standard ASR-33 teletype transmits and receives at 110 baud. Generally the same baud rate is used for both directions in a serial communication link, but there is nothing to prevent different rates from being used.

In *full-duplex* operation, transmissions in both directions are independent and simultaneous. A character transmitted to the computer by a terminal is not automatically printed or displayed; it must be "echoed" by the computer. In *half-duplex* operation, the terminal automatically prints or displays all characters that it transmits; echoing is not required.

SERIAL SIGNALS

At any time a serial transmission line may be transmitting one of two signals called *mark* and *space*. A 1 bit of a transmitted character (in the ASCII code, for example) is sent as a mark, and a 0 bit is sent as a space. However, the actual signals that are used to represent mark and space may vary at different stages in a communication system, as shown in Table M6-1.

TABLE M6-1 Mark and Space Signals

Medium	Mark	Space
Logic	1	0
TTL positive logic	2.4 v.	0.4 v.
20 ma current loop	20 mA	0 mA
EIA RS-232C	< -3 v.	> +3 v.
FSK (high band)	2225 Hz	2025 Hz
FSK (low band)	1270 Hz	1070 Hz

When a computer system generates a serial bit stream, it will represent mark as a logical 1 and space as a logical 0. If the computer uses TTL positive logic, then mark is 2.4 volts and space is 0.4 volts. To communicate with teletypes or certain other terminals, it may be necessary to convert these levels to *20 mA current loop* signals. A 20 mA current-loop interface consists of a pair of wires between transmitter and receiveer in which a current flow of 20 mA represents a mark, and the absence of current represents a space.

Most terminals transmit and receive levels dictated by *EIA Standard RS-232C*. As shown in Table M6-1, a mark is any voltage between -3 and -25 volts, and a space is any voltage between +3 and +25 volts. Computers and terminals that use the RS-232C standard can be connected to each other directly. Signals following the RS-232C standard can also be connected to a *modem* for transmission over a telephone line. A typical modem transmits a mark as a certain frequency in the audio band and a space as a frequency 200 Hz lower. This type of transmission is called *frequency shift keying (FSK)*. As shown in Table M6-1, two frequency bands are typically used, one for each direction of transmission. Thus a single telephone link can be used for bidirectional communication.

ASYNCHRONOUS COMMUNICATION PROTOCOL

The asynchronous communication protocol used by computers and terminals is illustrated in Figure M6-1(a). When the transmitter is idle (no characters being sent), the line is maintained at a continuous mark state. The transmitter may initiate transmission at any time by sending a start bit, that is, by putting the line in the space state for one bit time. It then transmits the data bits, least significant first, followed by an optional even or odd parity bit and 1, 1-1/2, or 2 stop bits. A stop bit is always a mark; after sending the stop bit(s), the transmitter may maintain the mark state for an arbitrary time before sending the next start bit. For full-speed transmission, the start bit is sent immediately after the stop bit(s). The period of time from the beginning of the start bit to the end of the stop bits is called a *frame*.

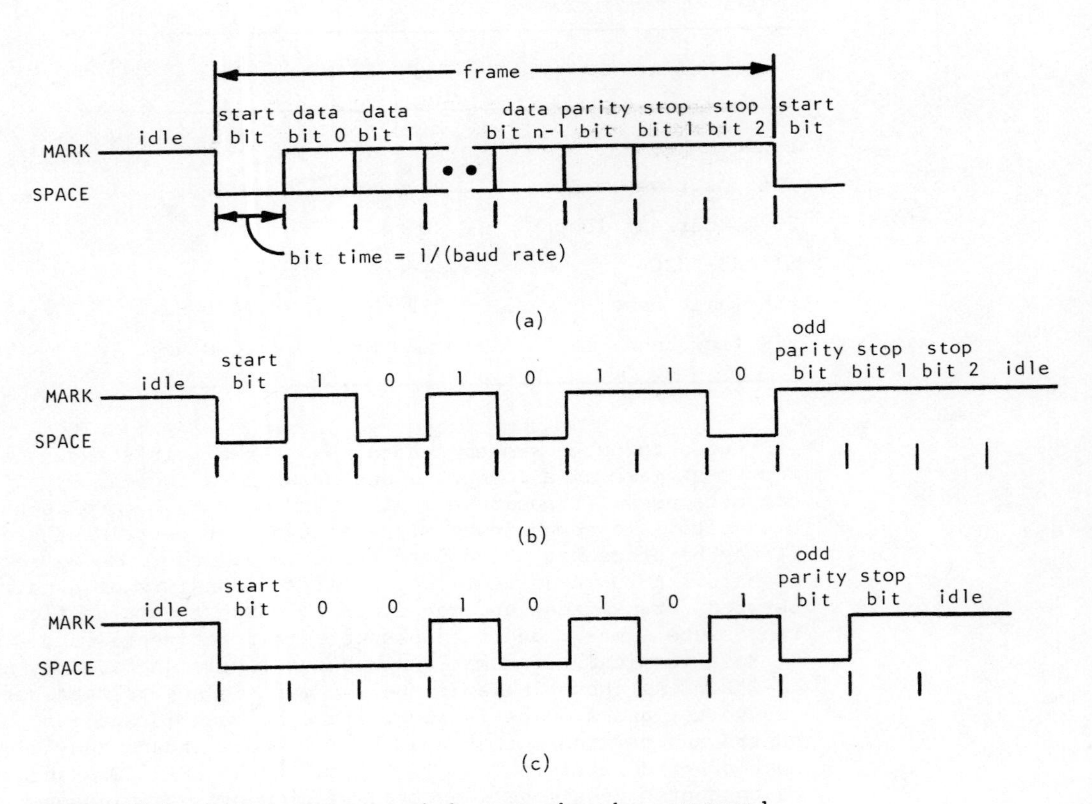

FIGURE M6-1 Asynchronous serial communication protocol.

A typical teletype operating at 110 baud transmits a start bit, 7 data bits, an odd parity bit, and two stop bits, as shown in Figure M6-1(b) for the character 0110101. Since 11 bits are required to send each character, at 110 baud there are 10 characters transmitted per second.

Terminals operating at 300 baud and higher usually send a start bit, 7 data bits, a parity bit with polarity that depends on the system, and one stop bit, as illustrated in Figure M6-1(c) for the character 1010100. Since only 10 bits are needed for each character, at 300 baud there are 30 characters transmitted per second.

DECODING SERIAL DATA

In a serial data communication system, the transmitter and receiver must agree on both the format and the baud rate of transmission. However, even if both ends agree on the baud rate, say 110 baud, it is impossible for both transmitter and receiver to have precisely matched frequency references. Fortunately, the protocol allows a decoding procedure, described below, that permits frequency references to differ by as much as a few percent.

The receiver of a serial bit stream should sample each bit in the middle of its bit time for optimum immunity to signal distortion and differences in receiver and transmitter frequency references. This can be accomplished by using a sampling frequency that is m times greater than the baud rate, where m is 16 or more. The decoding procedure, assuming the line is initially in the mark state, is as follows:

1) (Start bit detection) Sample the line for a space. After the first detection of a space, be sure that the line is in the space state for $m/2 - 1$ more samples. If successful, we are now approximately in the middle of a start bit. Otherwise we have detected a noise pulse and we should begin again.

2) (Data bit sampling) After m sample times, sample the line to obtain a data bit. Repeat n times for n data bits, and repeat if necessary for parity bit. Each bit will be sampled approximately in the middle of its bit time.

3) (Stop bit detection) After m sample times, sample the line for the mark state (stop bit). Repeat if there are two stop bits. If the line is in the space state, signal a "framing error." Otherwise, give the received character to the processor and go back to (1).

ASSIGNMENT

You are to build a microprocessor system that communicates with a terminal using full-duplex serial communication. The system should have a single bit input port for incoming serial data and a single bit output port for outgoing serial data. Incoming bit streams are to be decoded and outgoing streams are to be generated under program control. Timing is to be done by the use of wait loops or interrupts generated by a hardware timer.

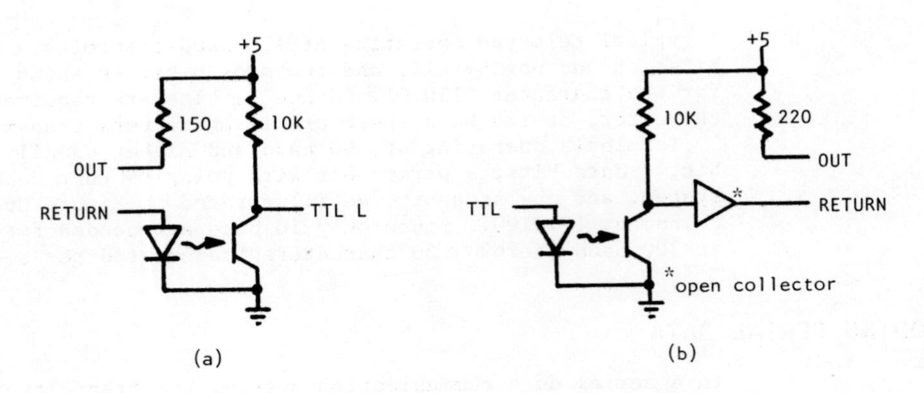

FIGURE M6-2 (a) Current-loop to TTL conversion; (b) TTL to current-loop conversion.

In addition to input and output routines, your program should have a simple monitor that allows the user to examine, modify and jump to locations in memory. You may devise your own format for monitor commands.

PARTS

You will need a microprocessor, basic system support chips, 256 bytes of PROM, and 128 bytes of RAM. Some microprocessors have a few bits of I/O ports built in. With other microprocessors, you will need separate I/O ports or a PIA for the two serial bit streams.

You will need circuits for converting the serial I/O from TTL levels to EIA or current loop signals. Commercially available chips such as SN75188, MC1488, and 8T15 perform the conversion from TTL to EIA levels, and SN75189, MC1489, and 8T16 convert from EIA to TTL. For TTL/current-loop conversion, opto-isolators can be used as shown in Figure M6-2. The circuits will work with a standard ASR-33 teletype, in which the serial output is produced by a simple switch opening and closing, and the serial input is received by a solenoid activated by a 20 mA current.

M7 - DIGITAL TACHOMETER

SUMMARY

This project uses the interrupt line of a microprocessor to sense the point firings of an automobile engine. A programmed counter that runs between interrupts measures the engine period, and a short program can then calculate and display the engine revolutions per minute (RPM).

INTRODUCTION

Principles of digital tachometer operation were discussed in Project C7. The need for frequent display updating and accuracy at low frequencies required an approach in which the period between point firings was measured and then inverted to find the frequency.

A microprocessor can be used to measure the period by using the leading edge of an engine point pulse to signal an interrupt. Initially a counter in a processor register or memory is set to zero and the interrupt system is enabled. The first interrupt transfers program control to a loop in which the counter is continuously incremented. The next interrupt transfers control out of the loop; the number in the counter is now proportional to the engine period.

Once the period is known it is a simple matter to obtain the RPM by disabling interrupts and dividing the period into an appropriate constant. The division can be carried out by a shift-and-subtract algorithm or by successive subtraction. (Would the brute-force counting method of Project C7 be sufficiently fast on a microprocessor?) The quotient can then be converted to decimal and displayed to an appropriate number of digits of accuracy (at least hundreds should be displayed). Then the interrupt system can be re-enabled and the process repeated. The time between display updates will be the greater of the time for division plus the engine period, and twice the engine period.

ASSIGNMENT

Design and built a digital tachometer based on a microprocessor. In the course of your design you should answer the following questions:

1) What is the time for execution of one iteration of your counter loop?

2) Based on the length of your counter loop, into what constant should the engine period be divided to obtain RPM?

3) What is the accuracy of your system?

PARTS

You will need a microprocessor, basic system support chips, and a 256-byte PROM. With some microprocessors you may be able to carry out all of the computations in registers, eliminating any need for RAM. Otherwise you will need one 128-byte RAM chip.

The only input to the system is a single line, the point pulses. The output consists of at least two 4-bit BCD digits (hundreds and thousands). These digits can be displayed in lamps or 7-segment displays. Input edge detection for interrupts and output latching can both be accomplished with a single PIA, or a flip-flop can be used for edge detection and 8-bit latches for output.

HINTS

When interrupts are re-enabled after the RPM calculating they should not respond until the *next* edge, that is, an edge occurring *during* the calculation should *not* be serviced.

Before the RPM calculation is begun, the period counter value should be initialized to reflect the overhead time (if any) spent between the edge of the first interrupt and the entering of the counter loop.

M8 - TURING MACHINE

SUMMARY

A microprocessor program can simulate the operation of a Turing machine. The Turing machine program is held in a PROM, and the Turing machine tape is simulated in RAM.

INTRODUCTION

As described in Project E5, a Turing machine has a control unit, a read/write head, and a scratch tape. The operation of the control unit can be described by a state/output table, or Turing machine program; a different program is used for each different Turing machine computation. In theory, a Turing machine has an infinitely long scratch tape, but many calculations can be performed with a finite tape simulated in RAM.

ASSIGNMENT

Design and build a Turing machine simulator based on a microprocessor. Your microprocessor program should interpret a Turing machine with a maximum of 64 states, including HALT. The program is stored as 128-byte table, one for each state/tape-input pair. A table byte has six bits to give the next state, one bit for tape direction, and one bit for the new tape symbol. The simulator should process a tape at a speed slow enough to be observed by a human user. Your simulator should have the following inputs and outputs.

TAPE - Seven lamps that show the tape cell currently positioned under the head and the three cells on either side of it.

MOVE - A lamps that blinks every time the tape is moved one cell in either direction.

RUN/PROG - A switch to determine the mode of operation. In PROG mode, the "tape" may be moved, inspected, and written by the user. When the switch is moved to the RUN position, the simulator begins operation in control unit state 1.

LEFT - A switch to move the tape left in PROG mode. The tape should move continuously (but slowly) while LEFT is activated.

RIGHT - Similar to LEFT, a switch to move the tape right.

COMP - A switch that, when operated, complements the contents of the cell under the head in PROG mode. Alternatively, provide two switches, one to specify a value to write and another to actually write it.

The Turing machine program should be stored in one 256-byte PROM and the microprocessor program in another. The tape can be stored in a 1K-bit RAM. Test your simulator with each of the programs specified in the assignment of Project E5.

PARTS

This project requires a microprocessor, basic system support chips, and two 256-byte PROMs. A 128-byte RAM can be used for program variables and for the Turing machine tape. If all program variables can all be stored in processor registers, then a 1K by 1 bit RAM can be used for the tape.

Eight bits of outputs and four or five bits of input are required. These can be provided with one PIA or an 8-bit output latch and a small input buffer. All switches should be debounced under program control, not by discrete flip-flops.

M9 - THE GAME OF DIB

SUMMARY

DIB is a game for two players. A microprocessor can be programmed to play DIB against a human opponent.

ASSIGNMENT

The game of DIB is described in Project D5. A machine can adopt an optimal strategy for playing DIB by referring to a state table (Table D5-1). You are to design and build a microprocessor system that plays DIB optimally against a human opponent. Your system should have the following inputs and outputs.

STICKS - Five switches to indicate the starting number of sticks, an odd number between 1 and 63.

LOAD - This switch loads STICKS and initializes a game.

USER - This switch is used to indicate the user's move. It is operated once for each stick the user wishes to take.

This switch is ignored after four pushes or if there are no sticks left.

MACHINE - This switch is used to tell the system that the user has finished taking sticks and is ready for the machine to move. Except at the beginning of the game, this switch is ignored unless USER has been operated at least once. (At the beginning of the game, either the machine or the user may go first, at the user's option.)

STICKSLEFT - Six lamps or two BCD digits to display the number of sticks left. The display is updated every time that USER is operated. When MACHINE is operated, the number of sticks taken by the machine should be flashed in the display, and then the remaining number of sticks should be left in the display.

WINNER - This lamp, off during the game, indicates the winner at the end of the game. It should flash quickly if the machine wins, slowly if the user wins.

PARTS

You will need a microprocessor, basic system support chips, and one 256-byte PROM. With most microprocssors no RAM will be needed - only one byte of storage for the number of sticks, a counter for wait loops, and a temporary register will be needed.

Seven bits or eight bits of output are needed. (If BCD is used, the highest bit of the highest digit is always 0.) Eight bits of input are needed. LOAD, USER, and MACHINE should be debounced by program.

M10 - MORSE CODE TO ASCII CONVERTER

SUMMARY

A Morse code transmission can be converted to a sequence of ASCII characters using a microprocessor program. The program dynamically adapts to varying transmission speeds.

INTRODUCTION

The Morse code was the first viable form of serial data communication, and for a number of years was employed in radio and land-based telegraph operations. Its main disadvantage is that it requires a skilled operator trained in the art of sending and receiving the code. For this reason, the Morse code has been superseded by voice and RTTY communication methods. However, Morse code is still used by approximately 250,000 amateur radio operators, as well as in long distance marine and aircraft communications. This project is aimed at developing a microcomputer system that can convert an incoming stream of Morse code characters into ASCII characters that can automatically be printed or displayed.

MORSE CODE

A Morse code transmission is a signal that at any moment is either on (1) or off (0). The transmission is decoded by observing both the sequence of transitions of the signal and the length of time in each state. The timing protocol is referenced to some fixed unit of time Δt:

dot:	signal at 1 for Δt
dash:	signal at 1 for $3\Delta t$
element space:	signal at 0 for Δt
character space:	signal at 0 for $3\Delta t$
word space:	signal at 0 for $7\Delta t$

A Morse code character consists of a sequence of one or more dots and dashes separated by element spaces. Characters in a transmission are separated by character spaces, and (English) words are separated by word spaces. Refer to ARRL [1976].

The time unit Δt depends on the transmission speed. Typical transmission speeds for a human operator range from 0.5 to 5 characters per second, with Δt ranging from about 0.2 to 0.01 seconds.

Of course, when Morse code is sent by a human operator, both Δt and the proportions of the elements referenced to Δt may vary over a small range. Nevertheless, it is still easy for a human to decode Morse code if the variances are not too large.

ASSIGNMENT

Design a microprocessor system that monitors an incoming serial bit stream and translates from Morse code into ASCII. Assuming that Δt of the incoming bit stream is no less than 20 ms (corresponding to 5 characters/second maximum speed), the incoming stream can easily be timed by a program loop. Each incoming 1 is determined to be a dot or dash and each incoming 0 is determined to be an element, character, or word space according to its length.

A reference value of Δt should be stored by the microprocessor. If an incoming 1 signal is shorter than $2\Delta t$ then it is assumed to be a dot, otherwise a dash. Element, character, and word spaces can be interpreted similarly by comparing the length of incoming 0 signals to $2\Delta t$ and $5\Delta t$. Varying transmission speeds can be handled adaptively. If a dot shorter than the reference Δt is received, then the reference is reduced slightly. Likewise if a dash longer than $3\Delta t$ is received, then the reference is increased slightly. If you wish you can experiment to find the optimum amount of the adjustment in terms of decoding accuracy and ability to track widely varying transmission speeds. A number of ideas can be found in articles in *QST*, January 1971 and October 1975.

Each time a complete Morse code character is received it should be translated into ASCII by means of a table look-up. You have to figure out how to carry out the table look-up. Then the ASCII character should be transmitted to a teleprinter or terminal. Your program should keep track of the number of characters transmitted and issue a carriage return and line feed before characters run off the end of each line.

Test your system by receiving Morse code transmissions. You will have to design a small circuit to convert the received signals to 1 and 0 levels compatible with the microprocessor inputs.

PARTS

This project requires a microprocessor, basic system support chips, one or two 256-byte PROMs, and 128 bytes of RAM. The only input is the Morse code signal, requiring a single input port or buffer. The ASCII terminal output can be provided through a UART such as the Intel 8251 or Motorola 6850.

M11 - DIGITAL STORAGE SCOPE

SUMMARY

A conventional oscilloscope can be used in conjunction with a microprocessor, memory, and DACs to create an oscilloscope that displays one-time signals indefinitely. Microprocessor control makes flexible triggering options easy to implement.

INTRODUCTION

Storage oscilloscopes are useful for observing one-time events and slowly changing signals that do not produce a readable trace on a conventional oscilloscope. Conventional storage scopes use special phosphors that retain a trace after it is written. A digital storage scope converts the input signal to a sequence of digital values, stores this sequence in a memory, and refreshes the screen of a conventional scope from the memory.

ASSIGNMENT

You are to design a microprocessor-based digital storage scope that displays input signals in a 256 by 256 dot matrix on a conventional X-Y display. (See PARTS for a method for using a triggered-sweep scope.) Figure M11-1 is a block diagram of one possible implementation of such a system. The sweep clock determines the rate at which the incoming analog signal is sampled, producing an interrupt for each sample time. The interrupt service routine converts the analog input signal to an 8-bit digital value, using a program controlled successive approximation algorithm as in Project M5. The converted value is stored in a 256-byte table in RAM; the display is blanked during the entire conversion by the Z control. Whenever the microprocessor is

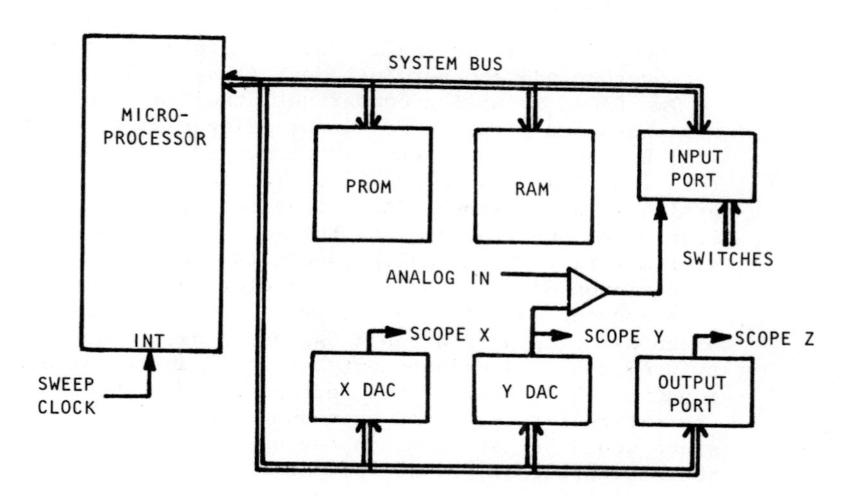

FIGURE M11-1 Digital storage scope block diagram.

not servicing interrupts it can refresh the scope display using the table in RAM. The program should go sequentially through the table, loading the table index into the X DAC and the table entry into the Y DAC, with the scope Z-axis enabled. Approximately 30 passes per second through the table will be required to produce a flicker-free display.

If input signals are sampled at the sweep rate and displayed continuously, then the system is said to have continuous or "free-running" sweep. However, microprocessor control makes possible many flexible triggering options. In addition to continuous sweep, your system should have at least a single sweep option, with a push button and a switch to select either positive or negative edge triggering. After START is depressed, the program should scan the incoming signal for a positive or negative edge as selected, and begin storing converted values and displaying them after the edge is detected. You may want to provide controls to vary both the sensitivity and voltage level of the triggering.

Another possible option is a form of delayed sweep -- the user selects a variable time delay between the triggering edge and the start of the display. An even more useful option, pretriggering, can be provided by implementing the table as circular queue. The queue is filled continuously, with overwriting as it wraps around, at the rate determined by the sweep clock. When the triggering event is detected, the display can be started at a user-selectable position in the queue -- from 0 to 255 sample times preceding the triggering event.

PARTS

This project requires a microprocessor, basic system support chips, and PROM. You will need 256 bytes of RAM for the table of converted values plus a small amount of RAM for program use. If the amount of scratchpad needed is small, you may prefer to reduce the table size rather than add another increment (128 or 256 bytes) or RAM.

Two DACs, an analog comparator, the Z-axis control, some user switches, and circuits for interfacing them to the microprocessor bus will be needed. Some DACs interface to a microprocessor bus directly (e.g., Analog Devices AD7522). You may wish to use a commercially available analog-to-digital converter rather than making a program-controlled one, to increase the maximum frequency at which your system works. Some analog-to-digital converters interface directly to a microprocessor bus (e.g., Analog Devices AD7570).

A conventional (triggered sweep) oscilloscope can be used instead of an X-Y display and the X DAC and Z-axis control can be eliminated by providing a single output bit to activate the external trigger input of the scope. For a refresh the microprocessor would activate the external trigger and then dump the table to the Y DAC. The scope speed would be set to match the speed of the dump. Using this method, the microprocessor cannot be interrupted during a refresh since the speed of the table dump must be uniform. Another alternative is to perform the dump under DMA control.

M12 - OSCILLOSCOPE PING-PONG

SUMMARY

The purpose of this project is to display on the screen of an X-Y oscilloscope a moving ball that bounces off the Ping-Pong paddles of two users. Features such as scorekeeping are easy to implement with microprocessor control.

ASSIGNMENT

This project is a microprocessor-based version of Project E2. As in Project E2, the most important function of the system is to display a moving ball on an X-Y oscilloscope. Associated with the moving ball are two position coordinates (PX and PY) and two velocity components (VX and VY) that are used to periodically update the position. The system must also keep track of and display the paddles and scores of two users.

In order to produce a flicker-free image the display should be refreshed 60 times per second. During each refresh period the following operations must be performed.

1) Compute new values of PX and PY by adding VX and VY, scaled appropriately.

2) Determine new paddle positions for both users.

3) If the ball is at the top or bottom edge of the screen, negate VY to "bounce" the ball. If the ball is at the left or right edge, determine whether it has hit a paddle. If it has, then

"bounce" the ball appropriately. Otherwise, increment the appropriate score and initiate a new serve after a short delay.

4) Display the ball.

5) Display the paddles.

6) Display the scores.

These operations are discussed in more detail below.

PX and PY should have approximately 14 bits of precision and VX and VY should be about 8 bits. The best format for PX, PY, VX, and VY varies with different microprocessors. Conceptually, the simplest format is to specify PX and PY as 14-bit unsigned (positive) quantities, and VX and VY as 8 bits of magnitude plus one bit for sign. Then PX and PY are updated as shown below for PX.

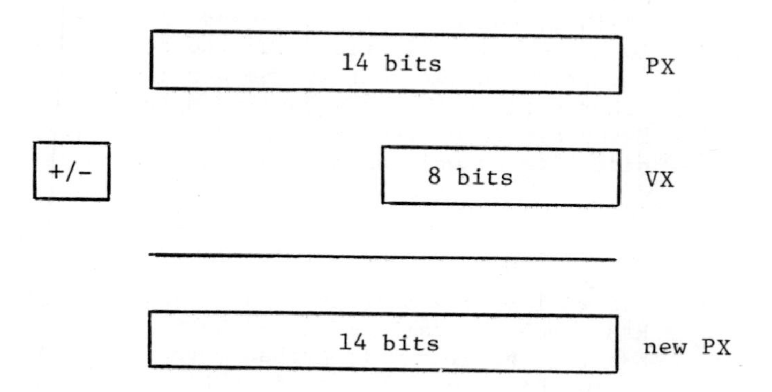

That is, the magnitude of VX is added to PX if the sign of VX is plus, subtracted if the sign is minus. The reader may wish to devise a format and corresponding operations where VX and PX are two's complement quantities.

The high-order 8 bits of PX and PY as defined above should be used to drive 8-bit DACs which in turn drive the corresponding scope axes. If VX or VY has its maximum magnitude then it will take 64 update steps for PX or PY to go from its minimum value to its maximum. Thus the shortest possible time for the ball to travel the screen is 64 · 1/60 second, or about one second. If VX and VY are smaller, the time will be longer.

When PY reaches its maximum or minimum value a bounce should be accomplished by negating VY. When PX reaches its maximum or minimum value a bounce takes place if the ball has hit a paddle. A paddle bounce should change the magnitudes of VX and VY as well as the sign of VX. The magnitude changes can be made randomly, or as a function of the position of the paddle hit (perhaps simulating a curved paddle) or the paddle movement (simulating english).

User control of the paddle may be accomplished in a number of ways. The microprocessor may maintain a position counter in memory for each paddle, incrementing and decrementing the counters at the users' push button commands, as in Project E2. Alternatively, potentiometers may be provided along with analog-to-digital converters to sense their positions.' The A/D converters can be implemented using the existing X and Y DACs and comparators under microprocessor control (as in Project M5), or separate units may be used. In any case, the displayed paddles should be 1/16 to 1/8 of the screen in height, and they should not be allowed to wrap around from the top to the bottom of the screen.

To perform the display function you will need a Z-axis control for the scope as well as X and Y DACs. The Z-axis is disabled while loading the X and Y DACs, and then enabled to display a dot. The length of time the Z-axis is enabled determines the relative intensity of the dot. The Z-axis is disabled whenever display operations are not being performed, including any time the DACs are being used for A/D conversion.

The scores should be displayed as two decimal digits per player at the top of the screen. A 5 × 7 dot matrix stored in memory may be used to generate each numeric character. To make the characters legible, you will probably need to expand the characters to a 10 × 14 ot 15 × 21 field for the scope display.

After you get your system working you may wish to experiment with other options that are easily implemented with microprocessor control. Examples are curved ball trajectories, paddles that move in the X direction as well as Y, and multiple balls.

PARTS

You will need a microprocessor, basic system support chips, PROM, and RAM. Only a small amount of RAM (128 bytes or less) will be needed, but your program will probably require 512 bytes of PROM or more.

Two DACs will be required to drive the scope X and Y axes. Some DACs such as the Analog Devices AD7522 have input latches and can be placed directly on a microprocessor bus, while others such as the Motorola MC1408L-8 will require the use of output latches or PIAs. One bit of output is needed for Z-axis control.

The only inputs required in the system are a reset push button and paddle controls. The two paddles can be controlled with four input switches or two potentiometers and A/D converters, as explained in the Assignment.

M13 - SCOPE-A-SKETCH

SUMMARY

This project is an application of direct memory access (DMA) in a microprocessor system. A bit map is stored in memory corresponding to positions in a 64 × 64 dot matrix. The bit map can be modified under program control by the microprocessor. At the same time, it is accessed by a DMA controller in order to refresh an X-Y display.

INTRODUCTION

Project D3 described a hard-wired system that could store a 64 × 64 dot matrix in memory and refresh an X-Y scope 60 times per second with the pattern. This requires a refresh data rate of about 250 Kbits/sec, too fast to implement under program control of a microprocessor. Nevertheless, the dot matrix can be stored and modified under microprocessor program control if direct memory access (DMA) is used for refresh. The refresh rate of 250 Kbits/sec translates to only about 30 Kbytes/sec. This means that a DMA controller needs only to fetch a memory byte every 33 μs, leaving the rest of the memory cycles available for microprocessor use.

ASSIGNMENT

You are to design a display system using a microprocessor and DMA as described above. The DMA should refresh the screen continuously from a fixed block of memory, at a rate sufficient to refresh the entire display 60 times per second. The microprocessor should clear the display memory upon reset, and then continuously scan for user commands and update the display memory accordingly. The cursor should be displayed as a blinking dot or square.

There should be two potentiometers for setting the X and Y positions of the cursor. The settings of the potentiometers should be determined by program-controlled analog-to-digital conversions. There should also be two switches or push buttons for specifying whether the dot at the cursor position is to be drawn, erased, complemented, or left unchanged, as in Project D3.

PARTS

You will need a microprocessor, basic system support chips and a 256-byte PROM for your program. You will also need 512 bytes of RAM for storage of the dot matrix. With some microprocessors you will be able to maintain the entire program state in registers; otherwise you will need another small RAM (128 bytes).

Two bits of input port will be needed for switches and two bits for analog comparators for the analog-to-digital conversions. A single DAC, driven by an 8-bit output port, will be needed (see Project M5). A single PIA is sufficient for all input/output ports.

The DMA can be implemented with MSI, or an LSI DMA controller such as the Intel 8257 can be used. A 12-bit counter will be needed to hold the scope X-Y position, and two DACs will be needed to drive the X and Y axes.

OPTIONS

Microprocessor control makes it possible to program many interesting features. One possibility is to change the system into a vector display by renaming the two push buttons "START VECTOR" and "END VECTOR." The switches are used in conjunction with the cursor to define the starting and ending points of vectors in the obvious manner. Each time END VECTOR is pushed, the program computes the positions of data between starting and ending points and "draws" the vector in memory. Additional switches can be added to allow the user to erase vectors, fill or erase the entire screen, and so on.

References

Advanced Micro Devices, 1975. *Am 2900 Bipolar Microprocessor Family*, Sunnyvale, Calif.

ARRL, 1976. *The Radio Amateur's Handbook*, 53rd Ed., American Radio Relay League, Newington, Conn.

Barna, A., and Porat, D. I., 1973. *Integrated Circuits in Digital Electronics*, John Wiley and Sons, New York.

Bartee, T. C., 1972. *Digital Computer Fundamentals*, 2nd Ed., McGraw-Hill Book Co., New York.

Bell, C. G., Grason, J., and Newell, A., 1972. *Designing Computers and Digital Systems*, Digital Press, Maynard, Mass.

Blakeslee, T. R., 1975. *Digital Design with Standard MSI and LSI*, John Wiley and Sons, New York.

Booth, T. L., 1971. *Digital Networks and Computer Systems*, John Wiley and Sons, New York.

Chaney, T. J., and Molnar, C. E., 1973. "Anomalous Behavior of Synchronizer and Arbiter Circuits," *IEEE Trans. Comput.* C-22, 421-23.

Digital Equipment Corporation, 1973. *RTM Register Transfer Modules*, Maynard, Mass.

Fairchild, 1972. *The Fairchild Semiconductor TTL Data Book*, Mountain View, Calif.

Fairchild, 1973. *The TTL Applications Handbook*, Mountain View, Calif.

Fairchild, 1975. *9400 Series Macrologic Composite Data Sheet*, Mountain View, Calif.

Gschwind, H., and McCluskey, E. J., 1975. *Design of Digital Computers*, 2nd Ed., Springer-Verlag, New York.

Hill, J., and Peterson, G. R., 1974. *Introduction to Switching Theory and Logic Design*, 2nd Ed., John Wiley and Sons, New York.

Hnatek, E. R., 1973. *A User's Handbook of Integrated Circuits*, John Wiley and Sons, New York.

Hnatek, E. R., 1975. *Applications of Linear Integrated Circuits*, John Wiley and Sons, New York.

Hopcroft, J. E., and Ullman, J. D., 1969. *Formal Languages and Their Relationship to Automata*, Addison-Wesley, Reading, Mass.

Intel, 1975. *8080 Microcomputer System User's Manual*, Santa Clara, Calif.

Klir, C. J., 1972. *Introduction to the Methodology of Switching Circuits*, Van Nostrand Reinhold, New York.

Kohavi, Z., 1970. *Switching and Finite Automata Theory*, McGraw-Hill Book Co., New York.

Kohonen, T., 1972. *Digital Circuits and Devices*, Prentice-Hall, Englewood Cliffs, N.J.

Mano, M., 1972. *Computer Logic Design*, Prentice-Hall, Englewood Cliffs, N.J.

McCluskey, E. J., 1965. *Introduction to the Theory of Switching Circuits*, McGraw-Hill Book Co., New York.

Melen, R., and Garland, H., 1975. *Understanding CMOS Integrated Circuits*, Howard W. Sams and Co., Inc., Indianapolis, Ind.

Minsky, M., 1967. *Computation: Finite and Infinite Machines*, Prentice-Hall, Inc., Englewood Cliffs, N.J.

Monolithic Memories, 1974. *6701 Data Sheets*, Sunnyvale, Calif.

Morris, R. L., and Miller, J. R., 1971. *Designing with TTL Integrated Circuits*, McGraw-Hill Book Co., New York.

Motorola, 1974. *McMOS Handbook*, Phoenix, Arizona.

Motorola 1975. *McMOS Integrated Circuits*, Phoenix, Arizona.

National, 1974. *CMOS Integrated Circuits*, Santa Clara, Calif.

Peatman, J. B., 1972. *The Design of Digital Systems*, McGraw-Hill Book Co., New York.

Peterson, W. W., and Weldon, E. J., 1972. *Error-Correcting Codes*, 2nd. Ed., MIT Press, Cambridge, Mass.

RCA, 1975. *CMOS Data Book*, RCA Solid State Division, Sommerville, N.J.

Signetics, 1974. *Signetics Digital, Linear, MOS Data Book*, Sunnyvale, Calif.

Sizer, T. R. H., 1968. *The Digital Differential Analyzer*, Chapman and Hall Ltd., London.

Stone, H. S., 1972. *Introduction to Computer Organization and Data Structures*, McGraw-Hill Book Co., New York.

Stone, H. S., 1973. *Discrete Mathematical Structures and Their Applications*, Science Research Associates, Inc., Chicago, Ill.

Texas Instruments, 1972. *The TTL Data Book for Design Engineers*, Dallas, Tex.

Tobey, G. E., Graeme, J. G., and Huelsman, L. P., 1971. *Operational Amplifiers Design and Applications*, McGraw-Hill Book Co., New York.

Turing, A. M., 1936. "On Computable Numbers with an Application to the Entsheidungsproblem," *Proc. London Math Soc.*, ser. 2, vol. 42, pp. 230-65.

Wait, J. V., Huelsman, L. P., and Korn, G. A., 1975. *Introduction to Operational Amplifier Theory and Applications*, McGraw-Hill Book Co., New York.

APPENDIX A Standard Kit of Integrated Circuits

5 - 7400 quadruple 2-input NAND gate

2 - 7402 quadruple 2-input NOR gate

2 - 7404 hex inverter

2 - 7410 triple 3-input NAND gate

2 - 7420 dual 4-input NAND gate

2 - 7474 dual positive-edge-triggered D flip-flop

4 - 7473, 7476, 74103, or 74107 dual negative edge-triggered J-K flip-flop

1 - 7483 4-bit adder

1 - 7486 quadruple 2-input exclusive-OR gate

2 - 7493 74161, 74163, 74177, 74191, 74193, or 74197 4-bit binary counter

2 - 7495, 74178, 74179, 74194, or 74195 4-bit parallel-access shift register

1 - 74151 8-line-to-1-line multiplexer

1 - 74153 dual 4-line-to-1-line multiplexer

1 - 74157 quadruple 2-line-to-1-line multiplexer

APPENDIX B

List of Integrated Circuits

1. SSI - 7400 series and 4000 series CMOS functional equivalents

7400 SERIES	4000 SERIES	DESCRIPTION
7400	4011	Quad 2-input NAND
7401	40107	Quad 2-input open collector NAND
7402	4001	Quad 2-input NOR
7403	40107	Quad 2-input open collector NAND
7404	4009, 4049	Hex inverter
7406		Hex inverter open-collector buffer
7407		Hex open-collector buffer
7408	4081	Quad 2-input AND
7409		Quad 2-input open-collector AND
7410	4023	Triple 3-input NAND
7411	4073	Triple 3-input NAND
7412		Triple 3-input open-collector NAND
7413		Dual 4-input NAND Schmitt-trigger
7414		Hex Schmitt-trigger inverter

7400 SERIES	4000 SERIES	DESCRIPTION
7415		Triple 3-input open-collector NAND
7416		Hex inverter open-collector buffer
7417		Hex open-collector buffer (high voltage)
7420	4012	Dual 4-input NAND
7421		Dual 4-input AND
7422		Dual 4-input open-collector NAND
7423		Dual 4-input NOR with strobe
7425	4002	Dual 4-input NOR with strobe
7426		Quad 2-input open-collector NAND high volt
7427	4025	Triple 3-input NOR
7428	4001	Quad 2-input NOR buffer
7430	4068	8-input NAND
7432	4071	Quad 2-input OR
7433		Quad 2-input open-collector NOR buffer
7437		Quad 2-input NAND buffer
7438		Quad 2-input open-collector NAND buffer
7440		Dual 4-input NAND buffer
7450	4085	Dual 2-wide 2-input AND-OR-INVERT
7451	4085	DUAL 2-wide 2-input AND-OR-INVERT
7452		4-wide AND-OR
7453	4086	4-wide AND-OR-INVERT
7454	4086	4-wide AND-OR-INVERT
7455		2-wide 4-input AND-OR-INVERT
7460		Dual 4-input expander
7461		Triple 3-input expander
7462		4-wide AND-OR expander
7464		4-wide AND-OR-INVERT
7465		4-wide open-collector AND-OR-INVERT
7470	4096	Gated J-K flip-flop
7472	4095	Gated J-K flip-flop
7473	4027	Dual J-K flip-flop
7474	4013	Dual positive-edge-triggered D flip-flop
7476	4027	Dual J-K flip-flop

7400 SERIES	4000 SERIES	DESCRIPTION
7478	4027	Dual J-K flip-flop
74101	4095	Gated negative-edge-triggered J-K flip-flop
74102	4095	Gated negative-edge-triggered J-K flip-flop
74103	4027	Dual negative-edge-triggered J-K flip-flop
74106	4027	Dual negative-edge-triggered J-K flip-flop
74107	4027	Dual J-K flip-flop
74108	4027	Dual negative-edge-triggered J-K flip-flop
74109	4027	Dual positive-edge-triggered J-K flip-flop
74110	4095	Gated J-K flip-flop
74111	4027	Dual J-K flip-flop
74112	4027	Dual negative-edge-triggered J-K flip-flop
74113	4027	Dual negative-edge-triggered J-K flip-flop
74114	4027	Dual negative-edge-triggered J-K flip-flop
74121	4047	Monostable multivibrator (one-shot)
74122	4047	Monostable multivibrator (one-shot)
74123	4098	Dual monostable multivibrator (one-shot)
74125	4502	Quad three-state buffer
74126	4502	Quad three-state buffer
74132	4093	Quad 2-input NAND Schmitt trigger
74279	4044	Quad $\overline{S}$-$\overline{R}$ latch

2. MSI - 7400 series and 4000 series CMOS equivalents

7400 SERIES	4000 SERIES	DESCRIPTION
7441		BCD-to-decimal decoder/driver
7442	4028	BCD-to-decimal decoder
7443		Excess-3-to-decimal decoder
7444		Excess-3-Gray-to-decimal decoder
7445		BCD-to-decimal decoder/driver
7446	4055, 4511	BCD-to-seven-segment decoder/driver
7447	4055, 4511	BCD-to-seven-segment decoder/driver

7400 SERIES	4000 SERIES	DESCRIPTION
7448	4055, 4511	BCD-to-seven-segment decoder/driver
7449	4055, 4511	BCD-to-seven-segment decoder/driver
7475	4042	Quad D latch
7477	4042	Quad D latch
7480		Gated full adder
7482		2-bit binary adder
7483	4008	4-bit binary adder
7485	4063	4-bit comparator
7486	4030, 4070	Quad 2-input EXCLUSIVE OR
7487		4-bit true/complement, zero/one element
7489		16 word × 4 bit RAM
	4036	4 word × 8 bit RAM
	4039	4 word × 8 bit RAM
7490	4518	Asynchronous BCD counter
7491	4015, 4094	8-bit serial-in, serial-out shift register
7492		Asynchronous divide-by-12 counter
7493	4520	Asynchronous 4-bit binary counter
7494	4035	4-bit parallel-in, serial-out shift register
7495	40104, 40194	4-bit parallel-in, parallel-out shift register
7496		5-bit parallel-in, parallel-out shift register
7497		6-bit binary rate multiplier
	4089	4-bit binary rate multiplier
74100	4034	8-bit D latch
74116		Dual 4-bit D-latch
74120		Dual pulse synchronizer/driver
74136		Quad 2-input open-collector EXCLUSIVE OR
74141	4028	BCD-to-decimal decoder/driver
74142		BCD counter/latch, decoder/driver
74143		BCD counter/latch, 7-segment decover/driver
74144		BCD counter/latch, 7-segment decoder/driver
74145	4028	BCD-to-decimal decoder/driver
74147		10-line-to-4-line priority encoder
74148	4532	8-line-to-3-line priority encoder

7400 SERIES	4000 SERIES	DESCRIPTION
74150	4067	16-line-to-1-line multiplexer
74151	4051, 4097	8-line-to-1-line multiplexer
74152	4051, 4097	8-line-to-1-line multiplexer
74153	4052	Dual 4-line-to-1-line multiplexer
74154	4515, 4515	4-line-to-16-line decoder
74155	4555, 4556	Dual 2-line-to-4-line decoder
74157	4019	Quad 2-line-to-1-line multiplexer
74158		Quad 2-line-to-1-line multiplexer
74159		4-line-to-16-line decoder
74160		BCD synchronous counter
74161		4-bit binary synchronous counter
74162		BCD synchronous counter
74163		4-bit binary synchronous counter
74164	4015	8-bit serial-in, parallel-out shift register
74165	4021	8-bit parallel-in, serial-out shift register
74166	4014	8-bit parallel-in, serial-out shift register
74167	4527	BCD rate multiplier
74170	4036, 3039	4 word × 4 bit RAM
74172		8 word × 2 bit multiport RAM
	40108	4 word × 4 bit multiport RAM
74173	4076	4-bit register with 3-state outputs
74174		6-bit register
74175		4-bit register with complementary outputs
74176	4518	Asynchronous BCD counter
74177	4520	Asynchronous 4-bit binary counter
74178	4520	4-bit parallel-in, parallel-out shift register
74179	4035	4-bit parallel-in, parallel-out shift register
74180	40101	9-bit parity generator
74181	40181	4-bit ALU
74182	40182	Lookahead carry generator
74190	4510	Synchronous BCD up/down counter
74191	4516	Synchronous 4-bit binary up/down counter

7400 SERIES	4000 SERIES	DESCRIPTION
74192		Synchronous BCD up/down counter
74193		Synchronous 4-bit binary up/down counter
74194	40104, 40194	4-bit parallel-in, parallel-out bi-directional shift register
74195	4035	4-bit parallel-in, parallel-out shift register
74196	4518	Asynchronous BCD counter
74197	4520	Asynchronous 4-bit binary counter
74198	4034	8-bit parallel-in, parallel-out bi-directional shift register
74199		8-bit parallel-in, parallel-out shift register

3. LSI - Microprocessors

PART #	MANUFACTURERS
8080	Intel, Texas Instruments (TI), Advanced Micro Devices (AMD)
6800	Motorola, American Microsystems Incorporated (AMI)
2650	Signetics, Advanced Memory Systems (AMS)
Z-80	Zilog
6502	MOS Technology
F8	Fairchild, Mostek
PPS-8	Rockwell
CP1600	General Instruments
TMS9900	TI
PACE	National

4. LSI - Memories

PART #	ORGANIZATION	MANUFACTURERS
1702	256 × 8 erasable MOS PROM	Intel, National
2704	512 × 8 erasable MOS PROM	Intel, National
2708	1024 × 8 erasable MOS PROM	Intel
8223,others	32 × 8 bipolar PROM	Signetics, National
74200,others	256 × 1 bipolar RAM	TI, Intel, others
2101,others	256 × 4 static MOS RAM	Intel, National, others
2102,others	1024 × 1 static MOS RAM	Intel, National, others
6810	128 × 8 static MOS RAM	Motorola
2107,others	4096 × 1 dynamic MOS RAM	Intel, Motorola, others
2116,others	16384 × 1 dynamic MOS RAM	Intel, others

5. MSI, LSI - Microprocessor support chips

PART #	DESCRIPTION	MANUFACTURERS
8224	8080 clock generator	Intel
8228	8080 system controller	Intel
6875	6800 clock generator	Motorola
8212	8-bit I/O latch	Intel
8T31	8-bit I/O latch	Signetics
6820	Peripheral Interface Adapter (PIA)	Motorola, AMI
8255	Programmable Peripheral Interface (PPI)	Intel
6850	Universal Asynchronous Receiver/ Transmitter (UART)	Motorola
8251	UART	Intel
8214	Interrupt control unit	Intel
6860	Low-speed modem	Motorola
8257	Programmable DMA controller	Intel
8259	Programmable interrupt controller	Intel
8253	Programmable interval timer	Intel
8216/8226	Bidirectional bus driver	Intel, Signetics
8T95/6/7/8	Three-state buffers/inverters	Signetics,others

6. Analog circuits

PART #	DESCRIPTION	MANUFACTURER
MC1408L-8	8-bit DAC	Motorola
DAC-IC8B	8-bit DAC	Datel Systems
DAC331	10-bit CMOS DAC	Hybrid Systems
AD7520	10-bit CMOS DAC	Analog Devices
AD7522	10-bit buffered CMOS DAC	Analog Devices
AD7570	10-bit ADC	Analog Devices
AD7506	16-channel multiplexer	Analog Devices

Index

Italics denote projects that stress the use of the indexed concept or component.